Advances in Geological Science

Studies in the twentieth century uncovered groundbreaking facts in geophysics and produced a radically new picture of the Earth's history. However, in some respects it also created more puzzles for the research community of the twenty-first century to tackle. This book series aims to present the state of the art of contemporary geological studies and offers the opportunity to discuss major open problems in geosciences and their phenomena. The main focus is on physical geological features such as geomorphology, petrology, sedimentology, geotectonics, volcanology, seismology, glaciology, and their environmental impacts. The monographs in the series, including multi-authored volumes, will examine prominent features of past events up to their current status, and possibly forecast some aspects of the foreseeable future. The guiding principle is that understanding the fundamentals and applied methodology of overlapping fields will be key to paving the way for the next generation.

More information about this series at http://www.springer.com/series/11723

Mitsuhiro Toriumi

Global Seismicity Dynamics and Data-Driven Science

Seismicity Modelling by Big Data Analytics

Springer

Mitsuhiro Toriumi
Japan Agency of Marine-Earth Science
and Technology
Yokohama, Kanagawa, Japan

ISSN 2524-3829 ISSN 2524-3837 (electronic)
Advances in Geological Science
ISBN 978-981-15-5108-6 ISBN 978-981-15-5109-3 (eBook)
https://doi.org/10.1007/978-981-15-5109-3

This Springer imprint is published by the registered company Springer Nature Singapore Pte Ltd.
The registered company address is: 152 Beach Road, #21-01/04 Gateway East, Singapore 189721, Singapore

Preface

Natural phenomena of the earth are still mysterious even if there are accumulating huge amount of observation data because of complex and diachronous process. To understand them, it may require to discover macroscopic invariances embedded in huge amounts of measured information in nature.

Earthquakes and volcanic eruptions are representative phenomena of solid earth dynamics, and they are often serious disaster for human society in the world. Though they occurred frequently in the world, scientists have not succeeded in earthquake prediction except for the case that the earthquake is associated with the obvious signatures before its event. Actually, almost always giant earthquakes occurred at the plate boundary, and intraplate large faults took place suddenly without any signals. In spite very wide regions around the giant earthquakes are damaged seriously not only on person but also society with huge amounts of infrastructure, it should be surprising that no signals before earthquake can be found in the recent investigations except for several rare examples.

Recent investigations on earthquake researches are based on the global networks of the precise wide-band digital seismometers, and on the foundation of dense network system seismometer stations in Japan at 1998 and USA after 1990, and later the ocean bottom geophysical stations network such as Dense Ocean Floor Network for Earthquake and Tsunami (DONET) system founded at 2011 in Japan and NEPTUNE of Canada at 2009. By these global and regional network systems established after 1980, the detailed structures of solid earth interior are possibly determined by the tomography technology from crust to deep mantle and core. As the results, the whole seismic velocity structure of the earth interior is inferred in detail though there are some inconsistencies in the seismic velocity structures in the mantle and crust in the active plate boundary. It may be due to the fact that the tomography images by present inversion method are based on the optimization with grid search technique but not global minimum search using Markov chain Monte Carlo method (MCMC). Recent investigation of seismic tomography has been developed with MCMC Bayesian inversion method for the synthetic data case. Then, near future should open the precise seismic tomography applicable to the real solid earth interior and give us the detailed seismic structure required for the determination of epicenters of the large to micro-earthquakes.

Global seismic networks are providing huge amounts of seismic data to scientists and person in various fields of the world, and due to this open sourcing of data, it is possible to carry the cross-checks by many different investigations that the proposed models on seismicity dynamics of earth interior are available for fundamental relations and predictions of seismicity both in the world and districts. Furthermore, it also provides the necessity for prediction where and what are important observation and seismic stations in the present districts and world. Huge amounts of monitoring and accumulating data should involve the fundamental earth mechanics laws in the long-term and short-term phenomena though these contain abundant noisy fluctuation. Today, the scientific tools for investigation of hidden fundamental processes governing the earth mechanics take methods for huge amounts of data and many different kinds of data in the real time scales by means of data-driven scientific methods involving machine learning and several regularization techniques and artificial intelligence. In addition, there are many techniques for data assimilation between the large-scale simulation experiments with huge number of parameters and observation big data in the present scientific environments. In the fields of global and regional climatology and oceanic dynamics, the integrated dynamics of ocean and atmosphere with cloud generation is well running on precise experimental equations governing the weather and climate in the world with deep learning technology and data assimilation programs using the monitoring data by satellite gravity (GRACE) and many oceanic buoys network system (ARGO). The solid earth mechanics field should be facing with the time to investigate the comprehensive studies on the global and regional seismicity dynamics based on the now accumulating monitoring data from the global and regional geoscience observation networks and scientific vehicles.

In this book, the author intends to introduce the global and regional seismicity dynamics based on the huge amounts of global and regional seismic source data possibly acquired from the world open databases by means of the data-driven sciences which should be a potential clue for the predictive investigation of the large earthquakes and mitigation of the giant disaster by them. This book may be useful for learning the recent and future data-driven scientific and machine learning methods applicable for many interdisciplinary studies of earth and planetary sciences.

In this book, the author intends to propose a new method based on machine learning of huge amount of accumulating seismic data both in the global and regional scales of solid earth, and thus, he think of importance of the open data sources made from established global and regional databases on seismicity by himself. Therefore, in order to access the database used here by readers, the author prepared the open database in Electronic Supplement Materials (ESM) in Springer.

Yokohama, Japan Mitsuhiro Toriumi

Contents

Chapter 1
Introduction

Abstract Since the last centuries, many giant earthquakes exceeding magnitude 9 associated with giant tsunami took place in the circum-Pacific coastal areas in the world. These natural disasters brought huge amounts of damages in our human communities. Most of the scientists deeply felt a necessity of fundamental understanding of global and regional mechanical behavior of earth process.

Keywords Giant earthquakes · GR law of earthquakes · Earthquake disaster

1.1 Introduction

In 2004 and 2011, the giant plate boundary earthquakes occurred in the wide ranges of Sumatra and Northeast Japan, and their magnitudes of the giant earthquakes reached over M 9.0 (Kanamori 2006; Fujiwara et al. 2011). The source faults of the Sumatra earthquake exceed 1000 km length and 20 m displacement along the Indian (Indo-Australia) plate (Satake and Atwater 2007), and the Tohoku-Oki giant earthquake ranges 500 km length and 300 km width along the plate boundary between the Pacific plate and the Japanese islands arc crust. The post-seismic displacement reaching 70 m at the edge of the overriding island arc crust displayed first eastward and then changed westward after one year, suggesting the recoupling of Pacific plate and island arc crust (Fujiwara et al. 2011). Giant plate boundary earthquakes exceeding M. 9 occurred six times during recent hundred years even around the Pacific Ocean: Kamchatka Peninsula earthquake, Alaska earthquake, Chile earthquake, Sumatra earthquake, and Tohoku-Oki earthquake (Fig. 1.1). The maximum slip extended about 1000 km along the Chile trench at the case of 1960 Chile earthquake but at the case of Tohoku-Oki earthquake, the scale of earthquake generating plate boundary zone reached 500 km × 300 km with the slip length of 70 m eastward.

Electronic supplementary material The online version of this chapter (https://doi.org/10.1007/978-981-15-5109-3_1) contains supplementary material, which is available to authorized users.

M. Toriumi, *Global Seismicity Dynamics and Data-Driven Science*, Advances in Geological Science,
https://doi.org/10.1007/978-981-15-5109-3_1

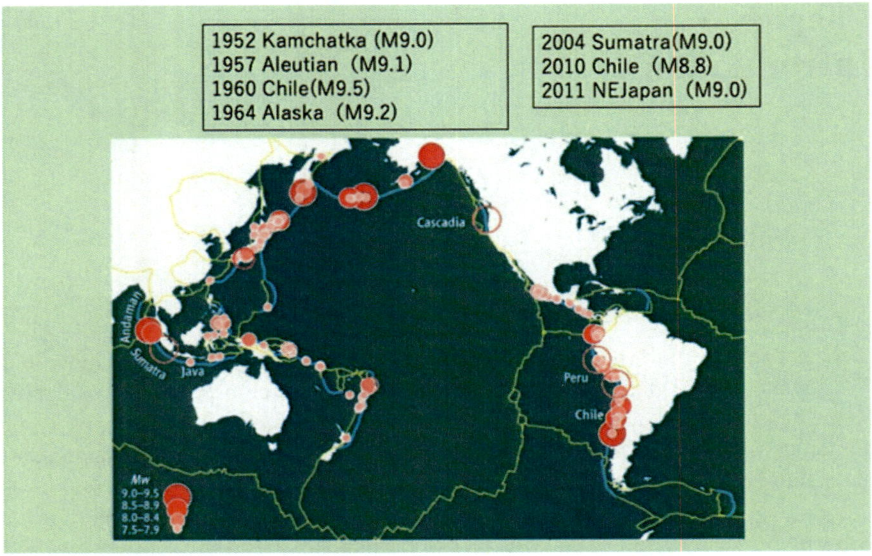

Fig. 1.1 Recent occurrence of giant earthquakes at the subduction zones around the Pacific Ocean from 1900 to 2016. (Personal communications of Dr. Sugioka of Kobe University)

Huge amount of damage by these giant disasters took a strong negative impact not only on regional community but also worldwide economy and social system because of destructive deficient effects about social system and social capitals together with many persons failed. In the case of Sumatra 2004, the plate boundary giant earthquake brought much wide disaster against the many coastal cities and towns around the Indian Sea from eastern Indian Sea even to the western area of the Indian Sea. At the same time, the giant tsunami with 10–30 m height generated from the Sumatra earthquake and attacked the coastal areas around the Indian Sea and persons over 300,000 died. In the case of Tohoku-Oki 2011 earthquake, very large tsunami reaching 30 m high also attacked the coast areas of Pacific side of the Tohoku District and persons over 25,000 died and lost.

As noted in many literatures, there were abundant giant earthquakes along the plate subduction boundaries of our planet which gave the great negative effects on human community and society. And it often gave a heavy damage-inducing the perfect destruction of one city or town. At the time of the Tohoku-Oki giant earthquake, the giant tsunami attacked the many small towns along the Sanriku District, and then those small cities were perfectly destroyed by the giant tsunami.

Today, the giant earthquake is considered to be generated by high-speed slippage along the large scale thrust faults and many of them occurred at the active plate boundaries, especially at the subduction plate boundary. The earthquake at the plate boundary is responsible for the unstable slip in the wide range of slip velocity, slip area, slip distance, depth of slip plane, and slip angle with other many parameters such as water film along the boundary plane as pointed out by Scholz (1982). Recent detailed investigations of the earthquakes occurred in the plate boundary zone show a manifestation that there is a wide variation of timescales of the durations characterizing the earthquakes ranging from subseconds to years. The very short time durations are characteristic in the case of very small magnitudes, but in the case of long-term durations of slip events the earthquakes appear wide variations of the magnitude reaching M 9. The relations between durations of the earthquakes and their magnitudes are recently suggested to be classified into two types by Ide and Shelly (2007): One is normal (regular) earthquake and another is slow slip event and tremor types. In the normal type, the magnitude seems to be proportional to the logarithmic duration time of the earthquake, and thus, the giant earthquakes arise generally long-term oscillation of the ground.

The earthquake is the elastic oscillation within the solid and fluid earth (Aki and Richards 1980). Thus, the earthquake is observed as the elastic oscillation of the ground, of the air, and of the seawater. Although the oscillation of the air and water is the sound itself showing the elastic body wave in the case of solid earth, it results in the elastic shear wave and body wave against the advancing direction of the elastic wave. The body wave is the oscillation of compression and expansion of the solid and fluid medium of the earth but shear wave is the oscillation of shear displacement of the solid medium normal to the advancing direction of the wave, and it is called as optical mode because of the oscillation geometry.

Elastic wave in the solid materials is primarily generated by the change of slip velocity of rock masses along on the slip plane (Aki and Richards 1980). Stable slip makes the elastic wave but almost always at the beginning of the slip or at the cessation of the slip the elastic wave is propagated and the timescale of the unstable slip corresponds to the frequency of the elastic wave. In general case, the slip velocity increases from zero to a finite value and then decreases to zero, indicating the maximum velocity change at the meantime of the slip motion along on the discontinuous plane. Then the observed oscillation of elastic shear wave at the ground station displays the mode of double couple of mutually orthogonal shear waves because of no rotation condition of the rock masses around the shear plane. However, it should be noted that the elastic wave generated by the explosion or implosion of rock masses such as phase transition of minerals and fluid phases appears to be compression and dilation modes, but it should be transformed into random orientation of double couple shear wave mode.

The earthquakes generating points are called as the hypocenter, but it means some extent of the slip plane area, being correlated with the magnitudes of them. The intensity of the earthquake is expressed by the magnitude (M) or moment magnitude defined by the following equation;

$$M = log\,(moment\ of\ earthquake) = log\,(du\ G\ S) \tag{1.1}$$

in which du is the shear displacement, G is the shear modulus (rigidity) of the rock, and S is the surface area of the shear slip plane (Aki and Richards 1980). Thus, it follows that the intensity of the earthquake shows those of the fracture area and displacement of the slip. Therefore, the unit increase of the magnitude corresponds

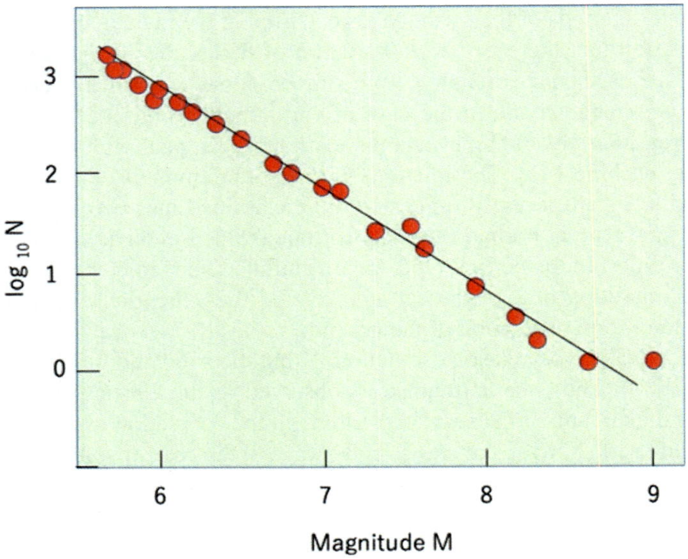

Fig. 1.2 Gutenberg and Richter law of the relation between magnitude and cumulative number of the earthquakes

to the ten times of the slip area in the case of constant slip displacement means 1/10 of that of earthquake magnitude. The total moment release by the M9 earthquake should be then 10^8 times of those of M1 earthquakes.

The logarithmic cumulative frequency of the earthquakes is proportional to the moment (i.e., the magnitude) as well-known Gutenberg–Richter law (Fig. 1.2),

$$N(M) = 10^{-bM} \tag{1.2}$$

This type of law is called as the scaling law, and the above GR law is satisfied in the very wide range of the seismic magnitude from M2 to M9. It is still strange in the physical mechanism of the very wide scale-free process in the natural solid earth mechanics as discussed by many authors (Allegre et al. 1982; Bak and Tang 1989; Ito and Matsuzaki 1990; Brown et al. 1991).

Combining these relations, it is obvious that in the large volume and time of earthquakes statistics the logarithmic cumulative frequency of the shear crack is proportional to the sizes of the shear cracks. Therefore, it seems that the very small shear cracks corresponding to M1 are very common in the crust and shallow mantle of the plate boundary zone, judging from that the 10^8 times of frequency compared with the M9 earthquakes. Therefore, if we consider the number density of micro-earthquakes below magnitude 2–3, the mechanical states and its time variation represented by density function of the very small shear cracks activated by small increase of the shear stress in the crust and mantle are possibly inferred in the global and regional solid earth. In addition, the mechanical state referring to the shear stress which is critical for propagating the shear cracks should be responsible for triggering the large giant earthquakes along the plate boundary and intraplate fault zones. In this book, the author would like to introduce the global seismicity dynamics by means of the characteristic features of the time series of density functions of micro-to-small earthquakes events in the global plate boundaries zones and highly seismic active subduction zone in the Japanese islands region and transform fault zone of Northern California region from 1990 to 2019 AD. The huge amounts of seismic data studied here are acquisited from seismic source data stored in the various databases in the USA (USGS, FDSN and Berkeley Seismicity Center and Japanese seismic open data centers (ERI of University of Tokyo, NIED, and JMA).

References

Aki K, Richards PG (1980) Quantitative seismology: theory and methods. W.H. Freeman, San Francisco, California, p 948

Allegre CJ, LeMouel JL, Provost A (1982) Scaling rules in rock fracture and possible implications for earthquake prediction. Nature 297:47–49

Bak P, Tang C (1989) Earthquakes as a self-organized critical phenomenon. J Geophys Res 94:15635–15637

Brown SR, Scholz CH, Rundle JB (1991) A simplified spring-block model of earthquakes. Geophys Res Lett 18:215–218

Fujiwara T, Kodaira S, No T, Kaiho Y, Takahashi N, Kaneda Y (2011) The 2011 Tohoku-Oki earthquake: displacement reaching the trench axis. Science 334:1240–1240. https://doi.org/10.1126/science.1211554

Ide S, Shelly DR (2007) A scaling law for slow earthquakes. Nature 447:76079. https://doi.org/10.1038/nature05780

Ito K, Matsuzaki M (1990) Earthquakes as self-organized critical phenomena. J Geophys Res 95:6853–6860

Kanamori H (2006) Lessons from the 2004 Sumatra-Andaman earthquake. Phil Trans R Soc A 364:1927–1945

Satake K, Atwater BF (2007) Long-term perspectives on giant earthquakes and tsunamis at subduction zones. Annu Rev Earth Planet Sci 35:349–374

Scholz CH (1982) Scaling laws for large earthquakes and consequences for physical models. Bull Seismol Soc Amer 72:1–14

Chapter 2
Nature of Earthquake in the Solid Earth

Abstract Earthquake is an oscillation of the solid and fluid of the earth interior, and it is excited by fracture of rocks. The modes of fracture and nature of medium, therefore, govern the propagation of elastic wave of earthquake, and thus, the nature of the earthquake is manifested by the global process of fracture in the earth interior.

Keywords Global distribution of earthquakes · Shear instability · Classification of earthquakes

2.1 Global Earthquake Distribution and Plate Tectonics

The plate tectonics has been founded by the global distribution patterns of earthquakes as discussed by Sykes (1967) and Isacks et al. (1968), together with magnetization anomaly of the ocean floor basalt as a kinematic model of rigid spherical plates along on the solid earth: The dense occurrence of many earthquakes appears in the narrow zone of the plate boundaries as shown in Fig. 2.1, and on the other hand, the rare occurrence of earthquakes apart from the plate boundaries except for large collision zone of the continents such as Tibet and southern Eurasia regions is as seen in the same figure (Fig. 2.1). The frequency of earthquakes along the mid-oceanic ridge zones, however, is lesser than those along the subduction boundaries (arc–trench). The ratio of the frequency between them reaches about 100–1000, and the maximum magnitude of the ridge zone is less than about 7 but that of the trench zone about 9.

The observations of earthquakes at many stations clarify the characteristics of the slip geometry of the shear cracks such as shear crack vectors composed of the shear plane and slip vector by the observed nodal planes by orientation of ground motions. The one of the shear cracks is the thrust type, second type is the transcurrent

Electronic supplementary material The online version of this chapter (https://doi.org/10.1007/978-981-15-5109-3_2) contains supplementary material, which is available to authorized users.

M. Toriumi, *Global Seismicity Dynamics and Data-Driven Science*, Advances in Geological Science,
https://doi.org/10.1007/978-981-15-5109-3_2

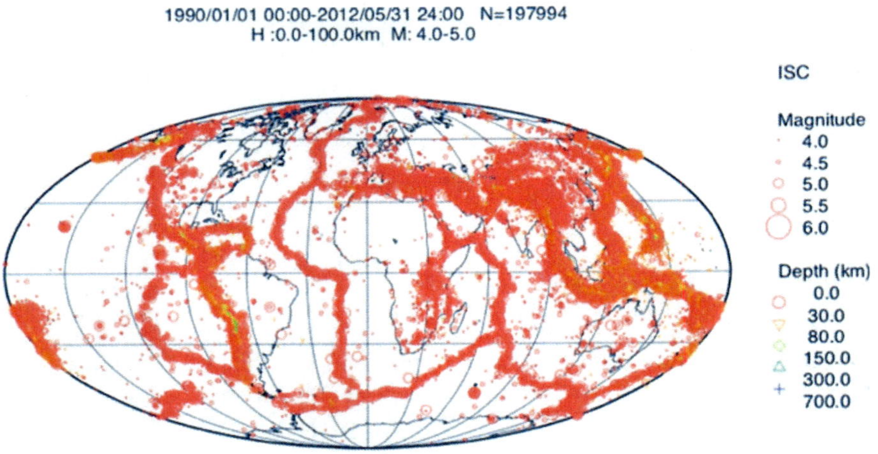

Fig. 2.1 Inhomogeneous distribution of earthquakes along global subduction zone and oceanic ridge zone

fault type, and third is the normal fault type. The difference of these three types is resulted from that of the stress geometry against gravity. The case of the maximum stress normal to the gravity and minimum one parallel to gravity is the thrust type, but the case of maximum stress parallel to the gravity and minimum one normal to the gravity is the normal fault type. Third case of transcurrent fault type corresponds to the intermediate stress parallel to the gravity. During the foundation stage of the plate tectonics from 1967 to 1970, it was stressed that modes of earthquakes in the subduction zone are characterized by the thrust-type and transcurrent-type faults, but in the ridge zone, the normal fault type is predominant.

The distribution patterns of earthquakes in the three-dimensional framework display sharply the oblique subduction of the oceanic plate behind the trench, which is called as Wadachi–Benioff zone from the ocean bottom level to 700 km depth. The characteristic frequency of earthquakes along the subducting oceanic plate is composed of the shallow maximum and the deep maximum as indicated by Sykes (1967) and as shown in Fig. 2.2. The former is called as shallow focused earthquakes and the latter as deep focused ones. The seismic mechanisms of both types have been investigated in the long time, and it seems that both are followed by double coupled type responsible for shear slip motion. However, the fracture mechanics suggests that the fracture strength becomes very large over several GPa in the 700 km depth, and shear cracks cannot be active by simple differential stress conditions under several MPa. Recent studies show that the deep focused earthquakes should occur by the rapid transformation of wadsleyite and ringwoodite after olivine incorporated with water in the 400–700 km depth (Green and Burnley 1989).

On the other hand, the seismicity of the ridge is characterized by the shallow earthquakes above 10 km depth, and almost always, it is represented commonly by

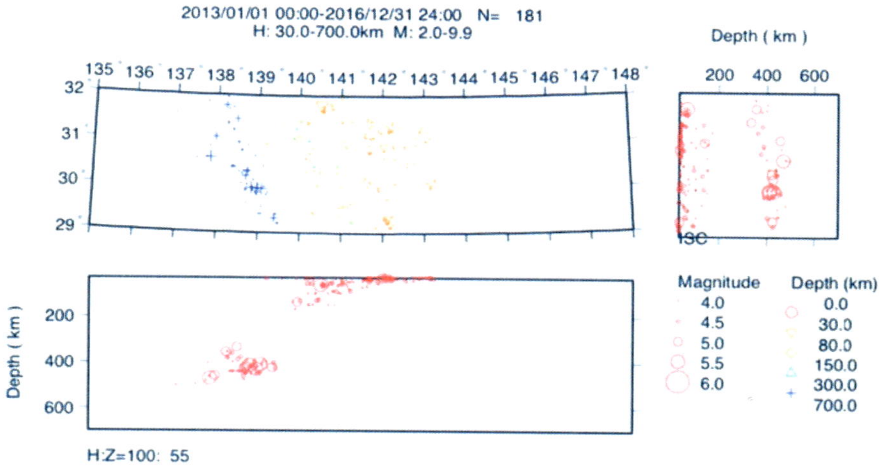

Fig. 2.2 Three-dimensional distribution of earthquakes under the Izu–Bonin arc, showing the two clusters of seismicity of the deep slab and shallow one of the Pacific plate

normal fault type. Such features are possibly considered to be due to high temperature conditions near the ridge and at large stress level should result in the yielding plastically at the relative shallower depth. Besides, the extensional stress field should be responsible for the low differential stress fracturing. Along the transform fault zones, the transcurrent fault type seismicity is common as manifestation of the plate tectonics kinematics (Isacks et al. 1968).

In summary, the global-scale earth mechanics is represented by seismicity patterns of the surficial solid earth with deep subduction zone at the narrow zones of the plate boundaries. Furthermore, this seismicity is caused by the pre-existing cracks growing associated with plate motion. The seismic wave as the earthquakes is emitted by the case that the shear cracks propagate with unstable mode indicating the shear instability in terms of solid mechanics (e.g., Rice 1993). In the next chapter, we will see the detail of the shear instability in the rock and solid mechanics resulting the earthquakes.

2.2 Earthquake Propagation and Shear Instability

Generally speaking, the seismic slip along shear cracks emitting elastic wave occurs unstably due to accumulating stress or strain in the solid earth. When slip in the solid earth takes place in the stationary constant speed, it leads stable slip, and it is responsible for permanent deformation by faults. Unstable slip phenomena of the pre-existing shear crack are resulted from the velocity softening and/or from the strain softening that is defined by the negative $\mathrm{d}F/\mathrm{d}V_s$ or $\mathrm{d}F/\mathrm{d}\varepsilon$ along on the slip zone, in which F is the force (shear stress) on slippage. Considering the frictional slip on the shear crack, the shear stress acting on the shear plane is followed by Coulomb–Navier law as,

$$\tau = \mu \sigma_n + c \tag{2.1}$$

Here, τ is the shear stress, μ is the frictional coefficient, and σ_n is the stress normal to the shear plane, respectively, and c is constant. The above criterion is the condition of frictional slip of the precut plane of the solid materials and rocks, but it does not suggest the unstable slip. The differentiation of right hand side that we obtain is in the case of unstable slip,

$$\mathrm{d}\tau/\mathrm{d}V = \mathrm{d}(\mu \sigma_n)/\mathrm{d}V = \mathrm{d}\mu/\mathrm{d}V < 0 \tag{2.2}$$

On the other hand, based on many experimental data of frictional slip of the rocks, Dietrich (1979, 1994) and Ruina (1983) obtained the rate and state equation of slip as follows;

$$\begin{aligned} \mu(\theta, V) &= \mu_o + a \ln(V/V_o) + b \ln(V_o\theta/D_c) \\ \mathrm{d}\theta/\mathrm{d}t &= 1 - V\theta/D_c \end{aligned} \tag{2.3}$$

where θ and μ are the state parameter and frictional coefficient of the slip, respectively, and V and D_c are the slip velocity and critical distance of slip, respectively. V_o is the normalized parameter of slip velocity.

In these equations, we obtain $\mathrm{d}\mu/\mathrm{d}V$ under the condition of $\mathrm{d}\theta/\mathrm{d}t = 0$ as follows

$$\mathrm{d}\mu/\mathrm{d}V = (a - b)/V$$

and

$$\tau = \sigma \mu \sim \sigma (a - b) \ln V \tag{2.4}$$

Therefore, the unstable condition of the slip is the case of negative value of $(a - b)$. The $(a - b)$ depends on the surrounding conditions such as pressure, temperature, fluid pressure, and other physicochemical parameters. Recently, Leeman et al. (2016) summarized the wide spectrum of slip instability of natural earthquakes phenomena from regular type to slow slip types as the marginal instability near the stable slip at $(a - b) \sim 0$. Unstable slip in the case of Eq. (2.4) should take place with softening by increasing slip velocity V as shown in Fig. 2.3.

The marginal instability near $(a - b) \sim 0$ then shows the instability of slips with various frequencies of critical slip distance (Ohnaka and Matsu-ura 2002), suggesting that the several long-term unstable slips may be coupled with each other and make some degree of resonance between them and start the unstable slip under the noisy fluctuations.

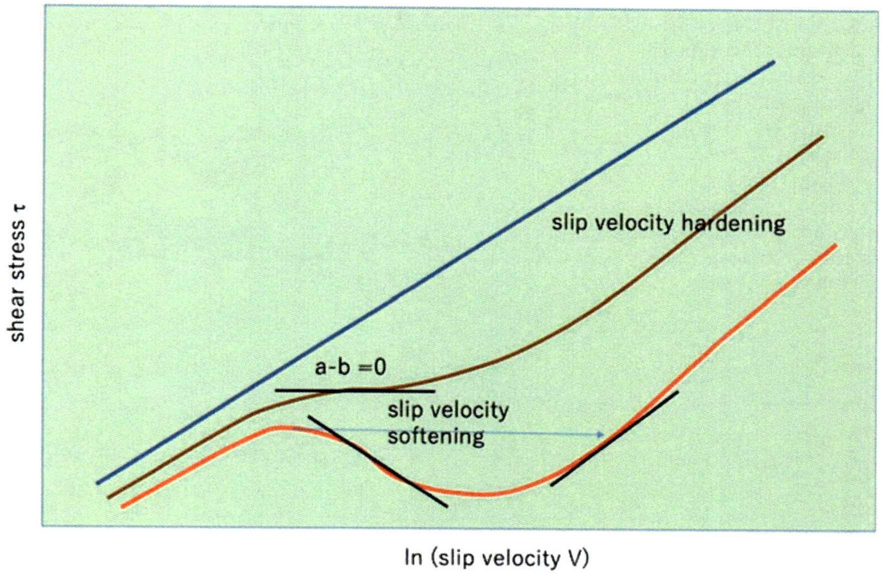

Fig. 2.3 Friction law of the shear slip plane with increasing normal stress from red to blue curve. The slip velocity softening appears in the negative slope of the curve and the hardening does in the positive slope regime. The critical condition at $a - b = 0$ occurs in the case of brown curve

By the way, we next consider the various types of network system of many slider block units as proposed by Burridge and Knopoff (1967) and later Brown et al. (1991), having the velocity v_i and displacement x_i as follows:

$$\begin{aligned} \mathrm{d}v_i/\mathrm{d}t &= k_i x_i + k_j x_{ij} + r_i v_i + \text{noises} \\ \mathrm{d}x_i/\mathrm{d}t &= v_i \end{aligned} \tag{2.5}$$

Here, k_i and r_i are the elastic constant and viscous resistance of the unit, respectively.

First, we consider the circular network system of simple slider blocks as shown in Fig. 2.4.

In order to investigate the network effects of multiple slider blocks connected by elastic and viscous stress such as the situation of the plate subduction zones around Pacific Ocean, the cases of the constant elastic modulus and viscosity and their fluctuated values are simulated. The system size is kept in the 100 units configured as the circular network of their connection. It follows that the block motion is characterized by the apparent random velocity and displacement, but the correlated motions of many slider blocks appear as the large-scale and long-term cooperative displacement along the circular connection as shown in Fig. 2.5. In this study, the correlated motions of the circular-connected slider blocks are processed by means of deterministic PCA method described in the later chapter.

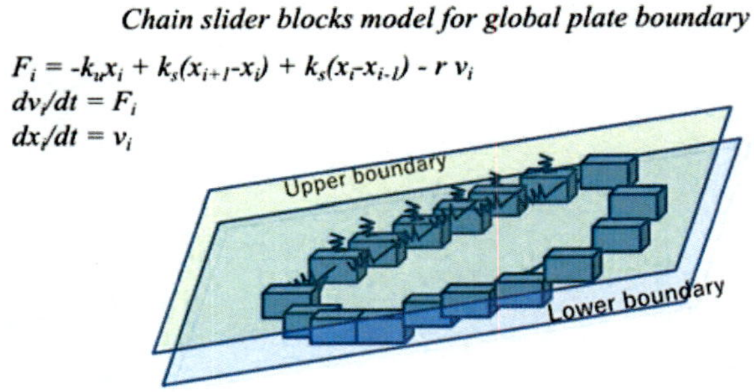

Fig. 2.4 Circular connected slider–spring block model showing the subduction zone slippages at the plate boundary. The upper boundary is of the subduction plate and overriding plate, and the lower one is with the asthenosphere underneath the subduction plate

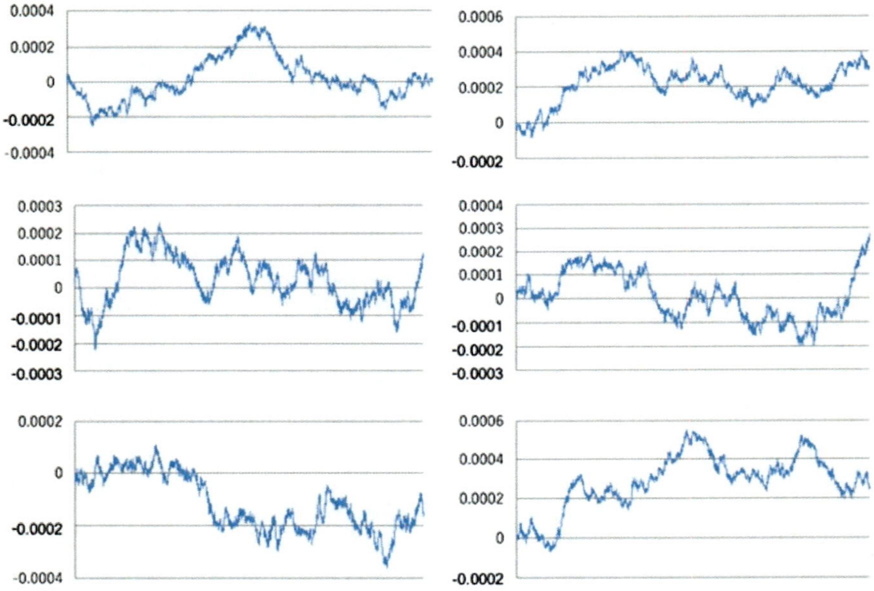

Fig. 2.5 Representative time series of the displacement of representative blocks (simulation; 5000 steps) on the circular connected slider block model. The vertical axis is the normalized displacement, and horizontal axis is the time step

On the other hand, the two-dimensional arrays of slider blocks shown in figures are simulated in the same parameter conditions as those of the chained slider blocks (Fig. 2.5). The correlated motions of the arrayed slider blocks can be seen clearly in the large-scale and long-term modes of combined individual blocks as shown in Fig. 2.6. By these simulation studies, clustered motions of the connected slider blocks are generated having the long-term oscillation modes.

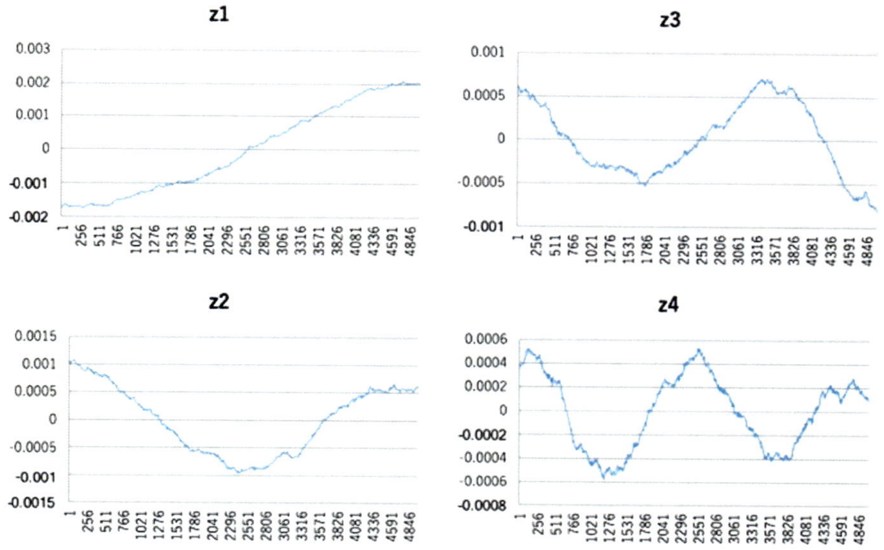

Fig. 2.6 Time series of major correlated displacements of z_1 to z_4 determined by principal component analysis of simulated blocks motion in the circular connected slider–spring block model of Fig. 2.4

2.3 Earthquakes and Global Network of Seismic Stations

In order to clarify the layered structure and the short- and long-term seismic activity of the solid earth, the global and regional digital seismometers have been founded in the world, and methods of automatic hypocenter determination by global networks are developed as CMT centroid moment tensor method by Harvard University group from 1980 (Dziewonski et al. 1981). The network stations are distributed as shown in Fig. 2.7, and global CMT hypocenters determined by these networks are stored in the global databases in the USGS earthquake center.

On the other hand, the Japanese regional networks of digital seismometers have been founded at 1998 after Hanshin–Awaji earthquake of 1995 as shown in Fig. 2.8. The database of seismicity in the Japanese islands regions is stored in database of JMA-1 by Japan Meteoritic Agency and National Institute of Earthquake Disasters. The database of seismic hypocenters involves earthquakes of M1 to M9 from 1998 to 2019. The area covers the hypocenters of Japanese islands regions from Okinawa to northern Hokkaido and Ogasawara islands. Recent ocean-bottom seismic networks of DONET I and II and also SNET are founded in the Off Kii and Shikoku regions of Nankai Trough and in the regions along the Japan Trench from Off Chiba to Off Hokkaido in order to monitor the near-field activity above Philippine Sea plate and Pacific plate, respectively. The DONET has the station nodes with seismometers and hydro-pressure gauges. The dense networks of sensitive hydro-pressure gauges can take the signals of the slow slip events and non-volcanic tremors along the plate

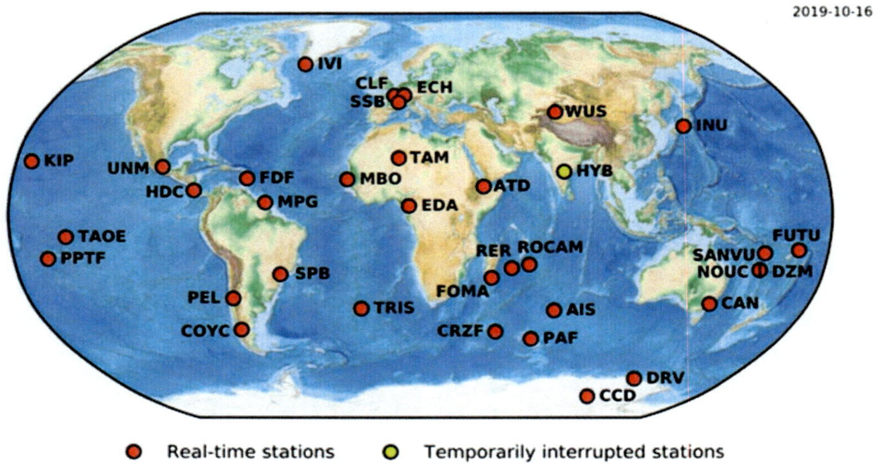

Fig. 2.7 Map showing the global network system of the seismicity of the solid earth cited from https://fdns.edu

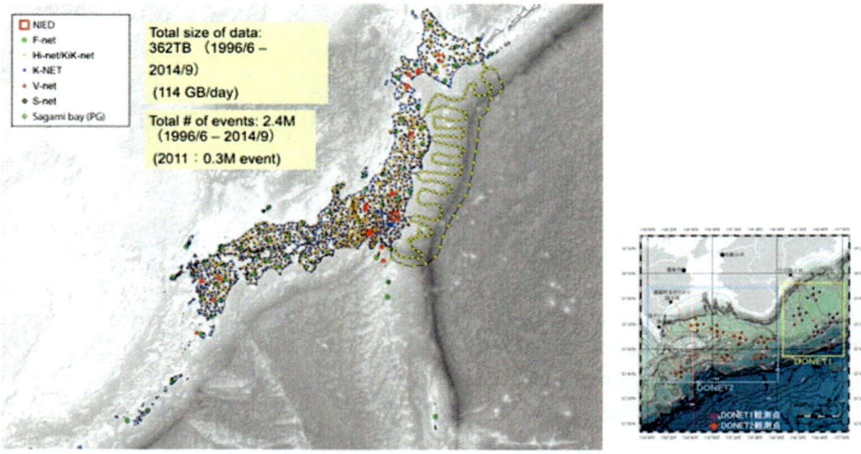

Fig. 2.8 Maps of regional network system of digital seismic stations and DONET ocean-bottom seismicity and tsunami observation system Off Kii and Shikoku in Japan (https://www.jamstec.go.jp/donet/j/donet/donet2.html)

subduction boundary and within the accretionary prism as being referred to Japan Agency of Marine Science and Technology (JAMSTEC) and National Institute of Earthquake Disaster (NIED) of Japan.

Regional networks of precise digital seismometers have been founded in various areas where large-scale transcurrent faults are active like the San Andreas Faults along west coast regions of US. The maps of the regional networks of seismic stations

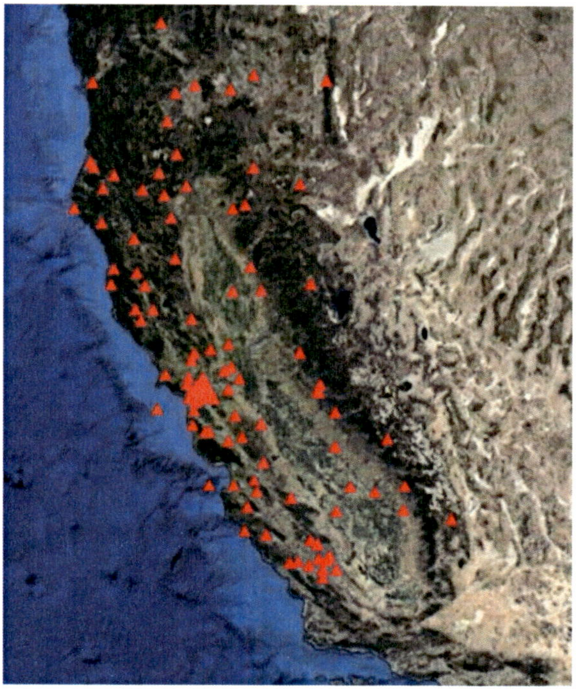

Fig. 2.9 Map of digital seismograph stations in Northern California cited from Northern California Earthquake Data Center, UC Berkeley Seismological Laboratory. Data set. https://doi.org/10.7932/ncedc: http://ncedc.org/berkeley-networks.html

are shown in Fig. 2.9, as being cited from International Federation of Digital Seismograph Networks (FDSN networks). These network databases are very useful for investigating the inner solid earth structures and various seismic activities related to the seasonal variations of rainfall and snow loading and unloading stresses and/or tidal force variation. The hydro-loading and unloading stresses can be evaluated precisely from the meteoritic databases of NOAA and from the gravitational variations observed from satellite observation systems (GRACE; Chanard et al. 2018). The data from the telemetric variations of continent—continent and hot spots—continent distances are also open accessed in database of VLBI centers of National Astronomical Observatory of Japan (ONAOJ; GSI home page (Geospatial Information Authority of Japan) 2019).

References

Brown SR, Scholz CH, Rundle JB (1991) A simplified spring-block model of earthquakes. Geophys Res Lett 18:215–218

Burridge R, Knopoff L (1967) Model and theoretical seismicity. Bull Seismol Soc Amer 57:341–371

Chanard K, Fleitout L, Calais E, Rebischung P, Avouac J-P (2018) Toward a global horizontal and vertical elastic load deformation model derived from GRACE and GNSS station position time series. J Geophys Res Solid Earth 123(4):3225–3237
Dieterich JH (1979) Modeling of rock friction, 1, Experimental results and constitutive equations. J Geophys Res 84:2161–2168
Dieterich J (1994) A constitutive law for rate of earthquake production and its application to earthquake clustering. J Geophys Res 99(B2):2601–2608
Dziewonski AM, Chou TA, Woodhouse JH (1981) Determination of earthquake source parameters from waveform data for studies of global and regional seismicity. J Geophys Res 86:2825–2853
Green HW, Burnley PC (1989) A new self-organizing mechanism for deep-focus earthquakes. Nature 341:733–737
Isacks B, Oliver J, Sykes LR (1968) Seismology and the new global tectonics. J Geophys Res 73(18):5855–5899
Leeman JR, Saffer DM, Scuderi MM, Marone C (2016) Laboratory observations of slow earthquakes and the spectrum of tectonic fault slip modes. Nat Commun 7:11104
Ohnaka M, Matsu'ura M (2002) The physics of earthquake generation. Univ. Tokyo. Press, p 378. in Japanese
Rice JR (1993) Spatio-temporal complexity of slip on a fault. J Geophys Res 98:9885–9907
Ruina AL (1983) Slip instability and state variable friction laws. J Geophys Res 77(88):10359–10370
Sykes LR (1967) Mechanism of earthquakes and nature of faulting on the mid-oceanic ridges. J Geophys Res 72(8):2131–2153

Chapter 3
Global Seismicity of the Solid Earth

Abstract Global seismicity of the earth is controlled by plate tectonics framework and then it is a key for clarifying the structure and dynamics of the earth interior. Especially, huge amounts of global seismic data are now accumulating and monitoring to be open in the scientific use involving the global mechanics of the short-term plate tectonics.

Keywords Seismicity of subduction zone and ridge · Distribution patterns of earthquakes · Percolation of seismic shear cracks

3.1 Stochastic Natures of Seismicity

Gutenberg and Richter (1944) have clarified the relationship between the magnitude and cumulative frequency of earthquakes in the logarithmic scales as shown in Fig. 3.1. That is called as GR law for seismicity and it holds as follows;

$$N(M) = k\,10^{-bM} \tag{3.1}$$

in which N is the total number of earthquakes larger than magnitude M, b is a constant to be called as b-value, and k is a constant.

The range of seismic magnitude satisfying this GR law appears from M2–3 to M7–8 without loss of observation, and the moment of earthquake must be in the huge wide range of seismic energy extending over 10^6. In the diagram of log N versus M, the nearly universal slope is defined as a b–value of GR law, and it seems that the population patterns of various seismic shear cracks active in the seismic region are unique even in the different time and region. In the seismicity under magnitude of 2, some degree of observation loss may be responsible for the deviation

---

Electronic supplementary material The online version of this chapter (https://doi.org/10.1007/978-981-15-5109-3_3) contains supplementary material, which is available to authorized users.

M. Toriumi, *Global Seismicity Dynamics and Data-Driven Science*, Advances in Geological Science,
https://doi.org/10.1007/978-981-15-5109-3_3

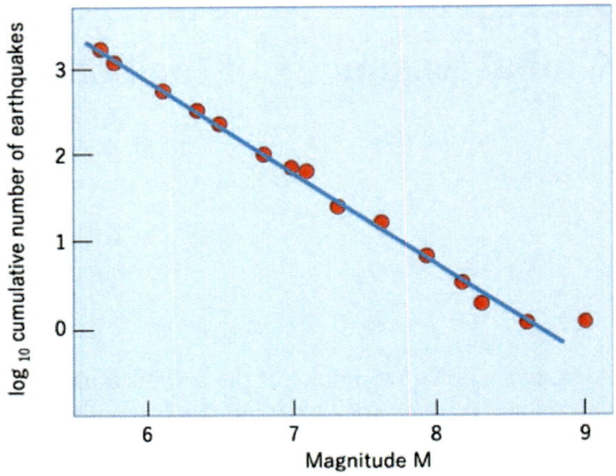

Fig. 3.1 Gutenberg–Richter law of magnitude and frequency of earthquakes in the world

of number of earthquakes from GR law, but recent progress in dense network of digital seismometers makes lowering the seismic loss in the many regions in the world. But the loss of seismic observations remains still in the global network system because of sparse distribution of seismometers in the oceanic regions.

Many recent studies of *b*-value in GR law have been performed in order to clarify the preseismic signals before large earthquakes and suggested that the *b*-value changes in time before and after the large earthquakes (Kanamori 1986). It means that the physical condition such as crack density and fluid pressure in the seismic region may change in time and space. Therefore, the author intends to discuss in the book such that the observation loss of seismic events smaller than magnitude 2 will be assumed as a constant coefficient dependent on the network observation system as discussed in the preceding sections.

As noted in the previous sections, the earthquakes are to be considered as shear cracks in the solid earth, and then the magnitude of the earthquakes emitted from active shear cracks should be approximated to be the relation of $M = \log (G\, S\, \mathrm{d}u)$ in which G is the rigidity of rocks, and S and du are the surface area and the displacement of the shear crack. Therefore, the product of S and du appears the power-law frequency. In many earthquakes, du is slightly dependent with the magnitude of earthquakes, and thus the size of shear cracks reveals power-law type frequency, and so the *b*-value of the GR law displays the abundance ratio of the shear crack size in the seismic region.

Judging from observations that the earthquakes of various magnitudes occur simultaneously in the same regions and in the long duration, it suggests that there are wide variations of shear crack sizes emitting earthquake wave at the same time and volume even in the short time duration.

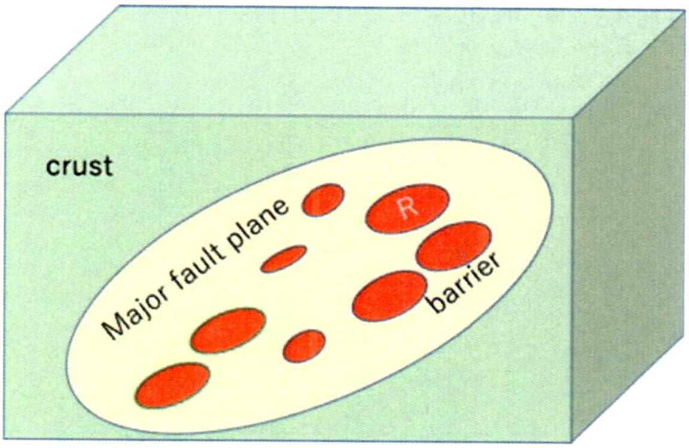

Fig. 3.2 In homogeneous distribution of the hard barriers with a various size R along the major fault plane

The similar barrier model for seismic slip of shear cracks was proposed by Aki (1979) and Kanamori and Stewart (1978) as asperity model: Aki (1984) considered the population of barriers on the large potential shear plane holds a power law of their size R as follows;

$$N(R) \sim R^{-d} \tag{3.2}$$

where d is the fractal dimension of the distribution of size R as shown in Fig. 3.2.

Using the relations of $N(M) = k\ 10^{(-d/2)M}$, we obtain

$$b = d/2 \tag{3.3}$$

Thus, the b-value of GR law is related to the fractal dimensions of population pattern of barriers, thereby suggesting the possible variations of b-value depending on the geological structures and their mechanical history of the seismic regions (Ohnaka and Mats'ura 2002).

Stochastic nature of the seismicity also appears in the epidemic occurrence of seismic events after large earthquakes (Fig. 3.3) that is formulated by Utsu (1961) as

$$n(t) = \mathrm{d}N/\mathrm{d}t = K/(t + t_o) \tag{3.4}$$

These types of aftershock seismicity populations are likely to the epidemic diffusion of seismic events just after the large earthquake as formulated later by Ogata (1988). Considering that there is a wide variation of the magnitude of the aftershock seismic events of which numbers follow above Omori's equation and that the b-value of these aftershock seismicity follows GR law, it is probable that the seismogenic shear crack barriers are distributed randomly and they decrease rapidly with time in the seismic volume.

The b-value variations have been observed in the stages before and after large earthquakes: The b-value becomes small before large earthquake and changes large after earthquake, but because of the drastic increase of aftershock seismicity the

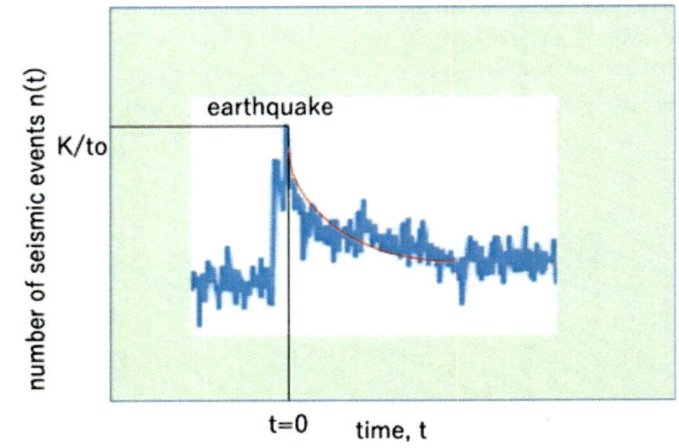

Fig. 3.3 Decay pattern of the seismic events after large earthquake following the asymptotic behavior of time

mode of change from small *b*-value to large one cannot be observed. However, many authors have a strong interest on the characteristic seismic signals before large and giant earthquakes near the plate boundaries mentioned above. But the *b*-value change before large earthquakes should be carefully checked by large number of seismicity trends before and after the large earthquakes, although *b*-values must be measured from probabilistic analyses by large number of earthquakes with various magnitudes. In order to estimate the precise *b*-values from many seismic events, the volume and the duration for counting the stochastic seismic events are required for rather large system of space and long duration. This type of trade-off relations between numbers of seismic events and sharp contrast of *b*-value change before large earthquake makes probably difficult to take available signals of *b*-value change.

3.2 Two Types of Earthquakes and Their Occurrences

Recent topics of the seismology are that there are two types of earthquakes in terms of event durations-magnitude relations found by Ide and Shelly (2007). The one is the regular type earthquakes that is defined by the slope is about 1/2 in the diagram of logarithmic duration of earthquake event and the magnitude as shown by Fig. 3.4. Another is the case of the slow earthquakes that display the slope of 1.0 in the same diagram. Both types vary the magnitude from M1 to M9 and the above linear relations are crossing around M2–M3 in the diagram described above. This type of simple relations between them is very enigmatic because the magnitude means the logarithmic moment and the duration is nearly the unstable slip time of earthquakes. As noted in the previous sections, the moment magnitude is expressed as

$$M = log\,(G\,S\,du) \tag{3.5}$$

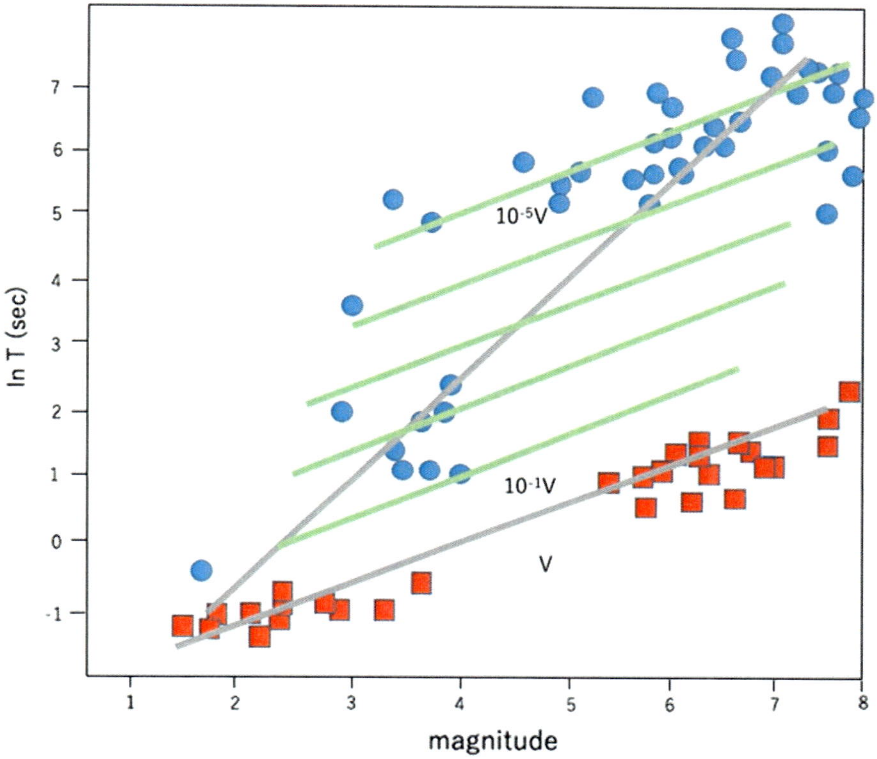

Fig. 3.4 Apparent relation between the magnitude and the duration of earthquakes. Two types of earthquakes (blue type is the slow earthquake and red one is the regular) are considered to be due to the difference of slip velocity of the earthquakes. V is the common slip velocity of regular type earthquakes, and tremor and slow slip event show 0.1 V and 10^{-5} V, respectively

and if the velocity of shear crack producing earthquakes is V, the duration (T) becomes

$$T\,V = \mathrm{d}u$$

and

$$S = S_0 + l(\mathrm{d}u)^2\left(\text{or } L(\mathrm{d}u)^1\right) = S_0\left(1 + l(\mathrm{d}u)^2/S_0\right) \sim S_0 \qquad (3.6)$$

in which l and L are the geometrical constant of expanding shear crack responsible for earthquakes.

On the other hand, S_0 holds (Kanamori and Stewart 1978),

$$\log S_0 = M - 4$$

Applying these relations to moment magnitude (Kanamori and Anderson 1975), we obtain

$$\log Mw = 1.5\,M + 9.1 = M - 4 + \log\,T + \log\,V + \text{constant} \tag{3.7}$$

and then

$$0.5\,M = \log\,T + \log\,V + \text{constant} \tag{3.8}$$

The former case corresponding to the regular earthquakes is the two-dimensional propagation of the shear crack and the latter corresponding to the slow earthquake is the one-dimensional propagation. The crossing points between them are considered to be the magnitude size of small seismic barriers proposed by Aki (1979), showing near the $M = 1 - 2$ as shown in Fig. 3.4. In this figure, there seems to exist another model for interpreting the relationship between the magnitude and the seismic duration. As shown in Eq. (3.4), the relationship between them contains the term of slip velocity V as,

$$\log\,T = 0.5\,M - \log\,V + \text{constant} \tag{3.9}$$

for regular earthquakes. In general, the velocity of shear slip of the earthquake ranges from several km per second for regular earthquakes to several to several km per month to year for slow slip events. Thus, the relations in Fig. 3.4 should be derived from the slip velocity systematics of the general earthquake but not from the two types of slip geometry mentioned above. At the case of this model, the relation between the duration and the magnitude of earthquakes should be normalized by the slip velocity and it seems the simple relation between them to be the same as the regular earthquakes as shown in Fig. 3.4.

The slow slip events called as SSE occur along on the boundary of the subducting slab in the depth from 10 to 40 km and show commonly periodic slip motions and various average slip durations ranging from several 10 s to several years. The periodicity of them also varies from several months to several to several 10 years even along the various plate subduction zones (Schwartz and Rokosky 2006). The relations above are schematically understood by the zonal structure of plate boundary composed of non-volcanic tremors, SSE, and megathrust zones (Ito et al. 2007). In this model, the non-volcanic tremor zones are divided into the deep and shallow zones and between the tremor zones there are SSE and megathrust zone which contains many barriers. These barriers correspond to those by Aki (1979) and Kanamori (1986).

These unstable shear crack slips are considered to be linked mechanically with each other in the subduction plate boundary extended along the trench axis, judging from their correlated periodicity as manifested in the circum-Pacific subduction regions (Schwartz and Rokosky 2006), though there are no obvious preseismic signals in NVTs and SSE before giant plate boundary earthquakes.

The megathrusts may involve the small many patches as barriers named by Aki and giant earthquakes may be responsible for massive slips of these barriers, being manifested by multiple large asperities as investigated by Kato et al. (2012) in the case of Tohoku-Oki giant earthquake (M9.0). Considering that the sporadic distribution of periodical repeated earthquakes (M4-M6) in the asperity regions of the Tohoku-Oki subduction zone (Igarashi et al. 2003) and the wide slip zone of Tohoku–Oki giant earthquake are mutually overlapped until 1990, the seismicity between the giant earthquakes is probably characterized by the stochastic fluctuation of the seismic activities of the intermediate and repeated earthquakes at the medium size barriers as suggested by Toriumi (2008).

Here, we think about the two modes of mechanical activity: The one is the inter-seismic mode I represented by multiple fluctuated activity of intermediate barrier slips and another is the co-seismic mode II represented by stochastic coherent activity of their barriers that is called as stochastic resonance of barriers slips on the single huge shear crack on the plate boundary. This model was first proposed by the author (ibid.).

Let us define the barriers having various sizes on the main slip plane as shown in Fig. 3.5. These barriers are mutually connected by elastic stress and connected by frictional stress with underneath main slip plane. Every barrier slips periodically but do not move coherently in the inter-seismic stage. In the co-seismic stage, we may think about the coherent slip of almost all barriers if these barriers are attracted mutually by entrainment excited by large fluctuations of the small noisy motions of main slip plane. In this situation, the barriers modeled by Aki (1979) may be considered as the repeated earthquakes by Igarashi et al. (2003), and unstable slip of them may occur in the manner of roughly periodical motion with a small elastic interaction between barriers (asynchronous mode). The large earthquakes should occur at the massive motion of their repeaters in the synchronous mode (synchronous mode). Thus, it seems that the non-synchronous mode changes rapidly to the synchronous mode at the time of the giant earthquakes large than magnitude 8. The transition from above two modes may be governed by the fluctuation of elastic strain around

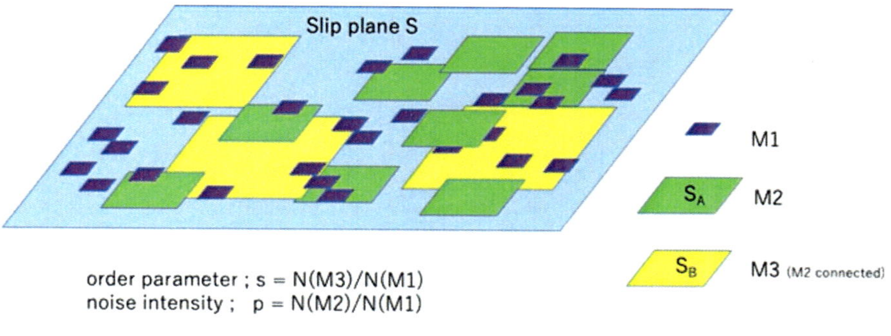

Fig. 3.5 Coagulation–percolation model of the different moment earthquakes for order parameter s controlled by noise intensity *p*

barriers and fracture stress lowering by water pressure increase, and others. These factors should induce the increase of seismic activity of micro to small earthquakes that is called as the seismic noise intensity in this book. This seismic noise may increase its intensity by the incremental slip motion by elastic interaction among repeaters even at the time of asynchronous mode. Then, the transition between these two modes should be considered as the change of the order parameter taking from 0 in the asynchronous mode and 1 in the synchronous mode.

Let us assume a coagulation enlargement of seismic slip area defining the coherent slips of the repeaters by defining the probability density of the seismic slip X_A and X_B (Fig. 3.5) as

$$X_A = N(A)S_A/S,$$
$$X_B = N(B)S_B/S, \tag{3.10}$$

in which S_A and S_B are the slip areas of magnitude *A* and *B*, respectively, and *S* is the seismic area considered here. *N*(*A*) and *N*(*B*) are the numbers of *A* and *B* units in the seismic area *S*.

Considering that coagulation of several seismic slip units of *A* makes seismic slip unit of *B* but seismic slip unit of *B* changes large seismic units due to coagulation with several seismic slip units of *A* in the manner of coagulation reaction as seen in Friedlander (1977) for aerosol and in Toriumi (1986) for flow aggregation of mineral grains in solid deformation as

$$\mathrm{d}X_B/\mathrm{d}t = k_1 X_A^n X_A - k_2 X_B X_A \tag{3.11}$$

in which k_1 and k_2 are the coagulation coefficients of *A* and *B* units with *A* unit, respectively. The term of X_A^n is assumed as the density of *n*-size cluster of *A* that becomes *B* by coagulation of single *A* unit.

Assuming the order parameter *s* and noise parameter *p* as,

$$X_B/X_0 = s$$
$$X_A/X_o = p \tag{3.12}$$

and, then

$$\mathrm{d}X_B/\mathrm{d}t = \mathrm{d}s/\mathrm{d}t + \mathrm{d}X_o/\mathrm{d}t$$

Let us assume $\mathrm{d}X_o/\mathrm{d}t \sim 0$, then it holds

$$ds/dt = X_A\left(X_A^n/X_o - s\right) \tag{3.13}$$

Now, considering that the coagulation coefficient should be proportional to the cross section of the slip unit $\mathrm{S}_{\mathrm{A-cluster}}^{1/2} \sim \mathrm{S}_{\mathrm{B}}^{1/2}$, above equation holds,

$$k_1 \sim k_2$$

Using the Eq. (3.13), we obtain,

$$\begin{aligned} s &= a - c\ \exp(-pX_o t) \\ a &= X_A^n / X_o \end{aligned} \tag{3.14}$$

here, c is the constant and it becomes $s = a - c = 0$ at $t = 0$.

Now let us define the order parameter s and noise paramer p as,

$$\begin{aligned} s &= N(M3 + M5 + .. + M9)/N(M1 + .. + M9) \\ &\sim N(M3)/N(MI) \sim 10^{-b'} \\ p &= N(M2)/N(M1 + M2) \\ &\sim N(M2)/N(M1) \sim 10^{-b''} \end{aligned} \tag{3.15}$$

in which $N(M)$ is number of earthquakes of magnitude M, and the noise intensity of the seismicity of the main slip plane. These parameters b' and b" are defined by the partial b-values from the GR law here, and then in the case of the large area and long time, the average of these partial b-values is nearly asymptotic to b of GR law.

Then, the Eq. (3.9) becomes,

$$s = \text{constant} \times X_A^n \tag{3.16}$$

here, it seems that $X_{M2} = S_{M2}\ (X_{M1}/S_{M1})\ p$ is approximated to p for $A = M2$. Then, it holds,

$$s \sim p^n \tag{3.17}$$

in which n is a parameter controlling the coagulation of seismic slip units to be about 9. This relation should be asymptotic solution at the steady state if there is a random fluctuation of p.

However, as denoted in the earlier sections, the b-value changes slightly with time and the range of seismic magnitude even in the inter-seismic and co-seismic stages, and thus the order parameter and noise magnitude display the different mode of the Gutenberg and Richter law between the larger earthquake and smaller earthquake clusters.

The b-values of GR law are possibly fluctuated by various parameters such as water content, micro-crack density, and metamorphic reaction, and thus it seems that above partial b-values satisfy;

$$b' = 2\,b + \varepsilon \quad and \quad b'' = b + \varepsilon'$$

in which ε and ε' are noise terms. Therefore, the following relation between the order parameter and seismic background holds,

$$s = k\ p^l \tag{3.18}$$

in which k is the rate of fluctuation of the b-values in the large and small magnitude region and l is a constant. This equation is similar to Eq. (3.17).

Using these parameters in the s and p diagram as shown in Fig. 3.6, we are possibly expressed the noise effect on the larger earthquakes on the subduction plate

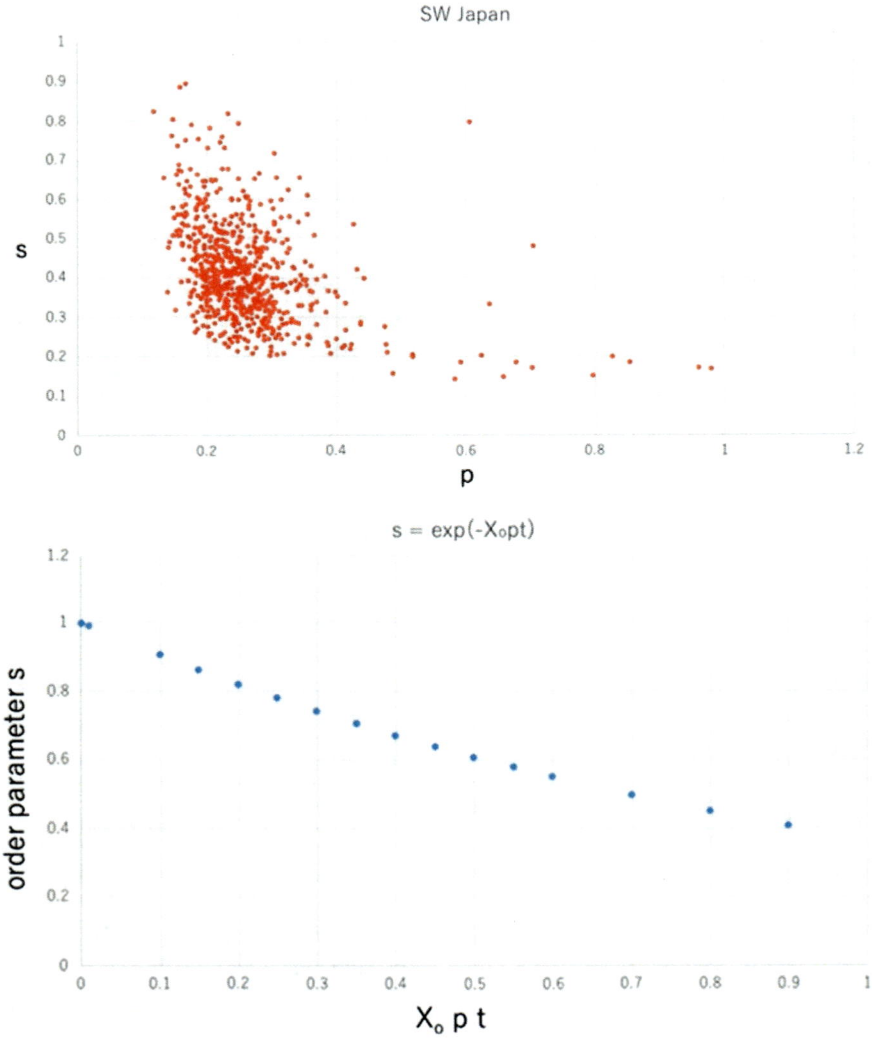

Fig. 3.6 Top—Relationship between the seismic order parameter s and noise p in the case of seismic activity in the Northeast Japan region from 1998 to 2018. The full lines indicate the relation of them obtained from the simple model in text: t_1 to t_3 is various time for model (top). **Bottom**—Relation between seismic order parameter s and noise parameter p in the simple kinetics model in text

boundary in the natural systems. The order parameter increases with increasing the noise parameter but the variance of the order parameter may show the peak value both in the cases of Tohoku-Oki fore-arc regions and SW Japan regions. The peak values of the order parameter variation are approximated to be about 0.2–0.3, suggesting the critical percolation of the barriers break out occupancy half of the geometrical critical percolation number (0.6–0.7). Therefore, it seems that the critical percolation of the barriers emitting the earthquakes should become smaller than that of the geometrical one because the actual size of the barriers become large for the stress concentration at their edges.

These discussions may allow us to imagine that the local *b*-values defined by partial slope of the GR law may change before and after the large earthquakes from small value to large one in the large magnitude region. On the other hand, in the small magnitude regions of GR law the local *b*-values should increase.

The stochastic scaling relation of co-seismic slip is rather commonly observed by many authors (e.g., Ohnaka and Matsu'ura 2002) though the constant co-seismic stress drops against the seismic magnitude (Ohnaka 2000). The scaling relations of co-seismic slip D of seismic fault and seismic fracture energy G become

$$G\left(J/m^2\right) = 10^{(2.4+1.4D(mm))} \quad (3.19)$$

The above empirical power law is available for the natural earthquakes and rock experiments (Ohnaka and Matsu'ura 2002).

It is important that the stress drops at the earthquakes are nearly constant around 10 MPa in the wide range of the seismic moment from 10^{13} to 10^{21} Nm. This is probably due to the constant yield stress of the barriers along on the main slip plane as mentioned previously and the number of barriers of large seismic slip plane are proportional to the surface area, that is, the density of barriers are constant. This is consistent with the stochastic resonance model of barriers on slip plane for the large earthquake described above (Toriumi 2008).

3.3 The Global Seismicity of Subduction Zones

The distribution of the hypocenters of the earthquakes is studied by many authors and as shown previously. Here, we will show the distribution patterns of deep focus earthquakes near the plate boundary, and the author intends to discuss the unstable slip phenomena emitting deep earthquakes.

The frequency of relatively large earthquakes shows rapid decrease with increasing depth, but it increases again in the deep subducting slab around the Pacific Ocean. Typical examples are obtained from the subduction zones of Tohoku-Oki, and Tonga-Kermadec, and Izu-Ogasawara from the seismic database of ERI using TSEIS cloud software. As shown in Fig. 3.6 of Pacific plate under the Izu-Ogasawara arc, it seems that the hypocenters of intraplate earthquakes display two clusters along the subducting slab: in the shallow region of 0–100 km cluster appears, and the deep seismic cluster over 400–500 km in the transition layer appears in that figure. The frequency distribution patterns of shallow and deep seismic zones display two segments composed of rapid decrease with increasing depth and maximum peak at 500 km depth as indicated by Oliver and Isacks (1967). It has been considered that shallower seismic activity above 200 km may be responsible for dehydration reaction enhanced softening of antigorite to olivine + H_2O with some amount of orthopyroxene and seismic peak at 500 km depth may be due to the coherent phase reaction from wadsleyite to ringwoodite with some amounts of Fe ion replaced with Mg ion (Ringwood 1975; Ito and Sato 1992). The driving force of shear crack propagations by these chemical reactions should be the strain accumulation and/or stress concentration by subducting oceanic plate. The simple model for seismic activity variations dependent on the fraction of reaction products such as serpentine to olivine + H_2O and wadsleyite to ringwoodite may be formulated by Avrami's equation (Avrami 1941) as

$$x = \left(1 - exp\left(-Kt^3\right)\right) \tag{3.20}$$

where x is volume fraction of reaction product ($0 < x < 1$), t is the time of reaction, and K is the geometrical constant in the adiabatic condition in the subduction process. The time t should be replaced by the distance L from the subduction point (the trench axis) using the plate velocity V_p as

$$L = L_o + V_p t \tag{3.21}$$

and thus, we obtain the following;

$$\begin{aligned} dx/dL &= (dx/dt)(dt/dL) \\ &= \left(3K(L - L_o)^2/V_p^3\right)exp\left(-K(L - L_o)^3/V_p^3\right) \end{aligned} \tag{3.22}$$

And, phase change induces the seismic events n as follows;

$$A\,dx = n\,S\,dL \tag{3.23}$$

in which A is a constant, and S is the unit area normal to the plate subduction as shown in Fig. 3.7.

Then, Eq. (3.15) becomes

$$n(L) = (A/S)\left(3K(L - L_o)^2/V_p^3\right)exp\left(-K(L - L_o)^3/V_p^3\right) \tag{3.24}$$

where V_p and L_o are the velocity of plate and start points of serpentine–olivine reaction (~50 km) and wadsleyite–ringwoodite transition (~500 km) as proposed by Green and Burnley (1989).

Figure 3.7 indicates the frequency distribution of seismic epicenters of M4 to M9 in the subducting slab, showing clearly that there are two peaks about 50 km and

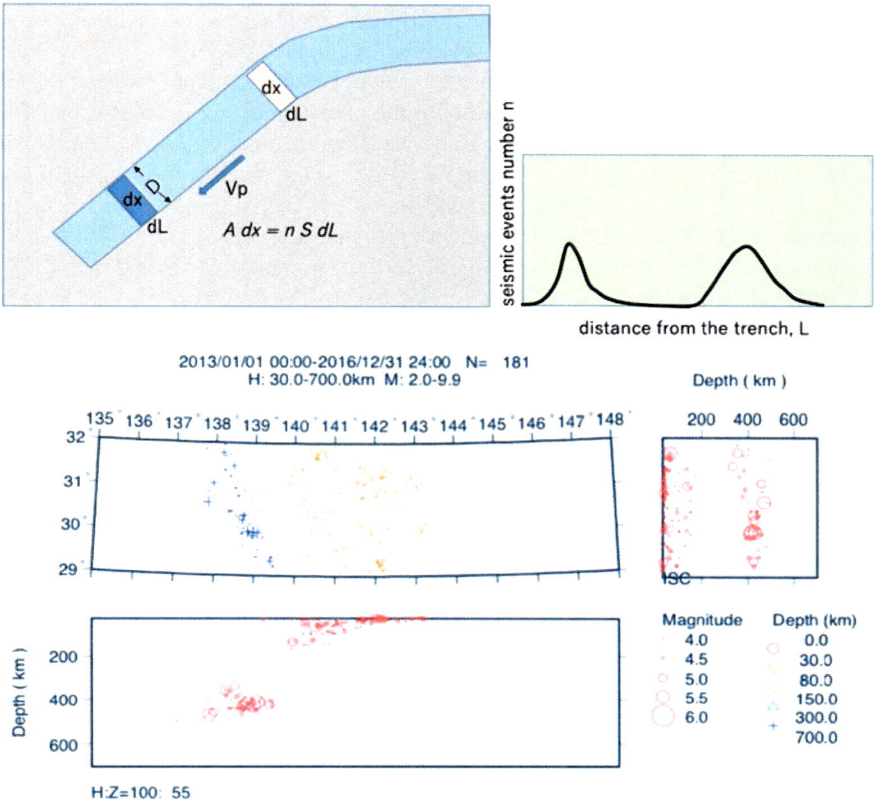

Fig. 3.7 **Top**—Model of double maxima of seismic frequency along the oceanic plate in terms of phase transition fraction x by Avrami's equation related to seismic activity n. In the incremental volume SdL, incremental volume fraction of new phase dx is proportional to the n in this model. **Bottom**—Three-dimension distribution pattern drawn by "TSEIS" of ERI in Univ. Tokyo of the seismic epicenters in the region of Izu-Ogasawara subduction zone of the Pacific plate. The distribution of along the subducted Pacific plate shows two clusters at depth shallower than 100 km and between 400 and 500 km

500 km. The probability curves of their seismicity are likely to the above production rate with plate depth from Avrami's equation.

On the other hand, the shallow seismicity having peak of 50 km appears some local clusters in the longitude–latitude map as shown in Fig. 3.7; the clusters are separated about 150–200 km within the subducting Pacific plate. If we can consider the dehydration model for the seismic activity having peak at 50–100 km, it seems that the hydrated regions are sporadically distributed in the slab. The size of the hydrated regions is possibly estimated as 100 km along NS direction and 200 km along the subducting slab, suggesting that the locally serpentinized (hydrated) slab was formed before the subduction of the plate as suggested by Omori et al. (2002).

It seems that the local hydration of the lithospheric mantle of the oceanic plate was performed in the outer-rise regions as clarified by Fujie et al. (2016).

On the other hand, in the deep intraslab earthquakes are frequent between 300 and 700 km depth, and these are responsible for the phase transition from wadsleyite to ringwoodite of Mg_2SiO_4. Considering that wadsleyite contains somewhat amounts of H_2O larger than olivine and ringwoodite, and thus the phase transition should be associated with liberation of H_2O to make possible to dehydration fracturing in the transition layer. During the slab subduction process, the serpentinized lithosphere should end the dehydration of serpentine and a little amount of H_2O (less than 0.1%) should be dissolved into surrounding olivine because of high solubility of H_2O in the high pressure conditions. The transition of wadsleyite from olivine may not be associated dehydration but H_2O may dissolve into wadsleyite because of high solubility (e.g., Karato 2008), suggesting no dehydration fracturing at this phase transition and scarce seismicity near the boundary between the upper mantle and the transition layer.

The deep earthquakes often show the large magnitude about M8, suggesting the large fracture area. This fact seems very strange because the width of fracture area may reach about 100–300 km and it is not reasonable to consider so much dehydration transition from wadsleyite to ringwoodite along the fracture area. However, it is probable that the positive feedback system operates in the overshooting conditions about the dehydration transition. It probably looks like the stochastic resonance of many seismic shear cracks in the shallow seismic regions discussed in the previous sections.

3.4 The Global Seismicity of Mid-Oceanic Ridges

The Mid-Atlantic Ridge and East Pacific Rise are the spreading centers of the oceanic plate and the spreading velocity of them becomes maximum at the positions 90 ° apart from the pole of rotation axes of the plate. However, the seismicity in the axial parts of the ridges does not show the large activity at the high spreading velocity zones. The activity of earthquakes in the global ridge zone is much smaller than that of subduction zones as shown in Fig. 3.8.

The seismic activity of the ridge is characterized by the normal fault type at the axial parts and by transcurrent fault type at the transform fault. The depth of the seismic activity is less than about 3–5 km and the magnitudes of the ridge earthquakes are commonly less than M5. It is noteworthy that the large earthquake (M7.1) occurred at the south of Southern Indian Ridge at 2015, and the focal mechanism was the normal fault type of which extension direction is parallel to the plate motion.

Near the ridge axis, the temperature in the mantle beneath the young thin plate is over 1000 °C, and thus the differential stress acting on the ridge axis should be over the plastic yield stress. Therefore, the rocks under the young thin plate should be deformed by plastic flow but not by fracture. The seismic activity at the ridge axis and transform faults is then ranged within the thin plate. The moment magnitude is

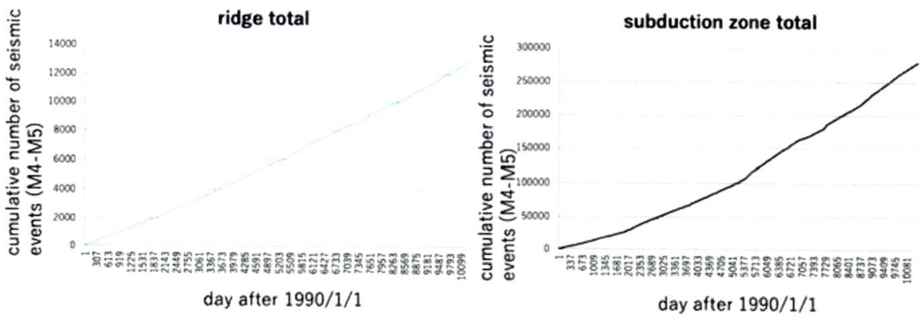

Fig. 3.8 Cumulative number of the seismic events of M4–M5 in the world from 1990 to 2018 in the global ridge and the global subduction zones

proportional to the fault area and so the limitation of the magnitude is responsible for the geometry of the normal fault size limited to be 5 km depth.

The global distribution of earthquakes along the oceanic ridge is displayed in Fig. 2.1, showing the linear arrays hypocenters and their trace crosses with the hot spot type oceanic island such as Iceland, Galapagos, and Easter. The volcanism of these oceanic islands is very active and beneath them there are large scale hot plume in the mantle by means of satellite gravity tomography and seismic tomography.

Recently, the large normal fault type earthquake occurred at the midpoint between Kerguren islands and south Indian ridge reaching the magnitude of 7.1. Judging from such large magnitude, the seismogenic normal fault has probably the fault plane reaching about 5 km depth and 20–30 km width. The size of fault plane is

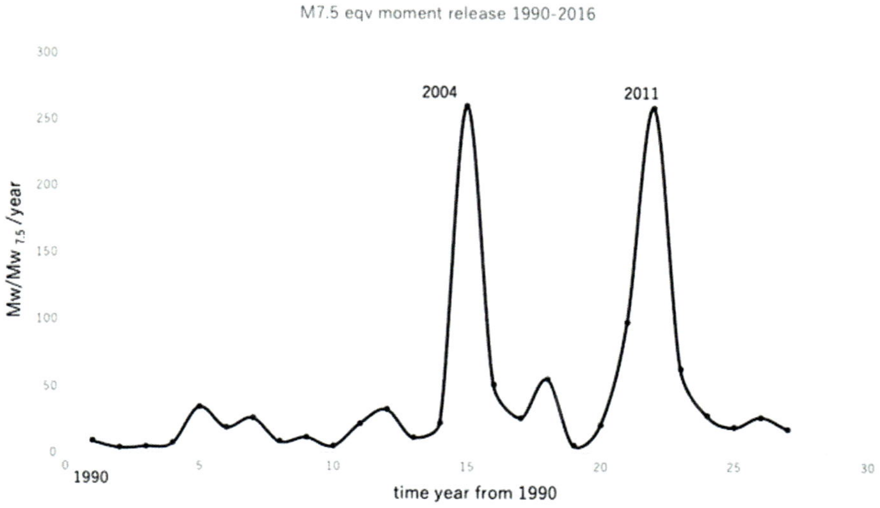

Fig. 3.9 Time series of global moment release rate per year normalized by M7.5 from 1990 to 2016 from data of USGS data center, showing the two large peaks at 2004 and 2011

abnormally large rather than earthquakes at the ridge axis, but it is noteworthy that intraplate earthquakes at south of Hawaii Island (1975) and west of the Sunda Island (2012) were about 7.4 and 8.6, respectively. The former was the normal fault type and the latter was the strike-slip fault type. Furthermore, normal fault type large earthquakes occurred at the outer-rise regions of Japan Trench (2011) and Izu-Mariana Trench (2010). Considering these occurrences of intraoceanic plate earthquakes, seismic activity within the oceanic plate is not scarce but it is enhanced by the active subduction and ridge seismicity because of mechanical continuity through the elastic plate.

3.5 Global Moment Release Rates by Large Earthquakes

Since the global network of digital seismic stations has been established and available method of global hypocenter and source parameters of the earthquakes have been developed by Dzivonski et al. (1981), it is possible to determine the global seismic moment release rate per year from 1990 to 2020. Ekstrom (2007) has tried to investigate the time series of the global moment release rate displaying the two peaks at 2004 when the Sumatra giant earthquake (M 9.2) occurred and at 2011 of Tohoku-Oki giant earthquake (M 9.0). The author tried to determine the moment release rate variation in terms of M 7.5 equivalent numbers per year in the world to detect the time series of total seismic moment release rate in the global scales, using the database of USGS seismicity archives. The time series of the rates of M7.5 equivalent number of earthquake is shown in Fig. 3.9, sharply displaying the two peaks at 2004 and 2011 and the general increase after 2003. The peaks are responsible for the very large earthquakes such as Sumatra (2004) and Tohoku-Oki (2011). Thus, we may infer the peaks at 1952, 1957, 1960, and 1964 before 2004 and 2011. It is noteworthy that there seems to be big moment release rate periods of 1950–1965 and 2004–2011.

On the other hand, the large and giant earthquakes larger than M 8.5 from 1500 AD to present are possibly obtained from the database in JMA (Japan Meteroic Agency) as shown in Fig. 3.10. It seems that the large seismic activity in the world has been changed repeating up and down: During 1500 to 1700 AD and from 1800 to 2000 AD, the seismic activity around the Pacific Ocean was very strong, but the period between 1700 and 1800 AD it is not so large. Furthermore, the figure indicates that the large seismic activity along the subduction zones of the South America including Peru and Chile is very active but in the northern Pacific Ocean and Indonesian regions of the subduction zones the large seismic activity becomes recently large enough compared with that of the period before 1900 AD. Furthermore, it should be noted that giant earthquakes from 1500 to 2016 AD become very frequent in the narrow time intervals as shown in Fig. 3.11, and thus the giant earthquakes occur as like as a cluster in time.

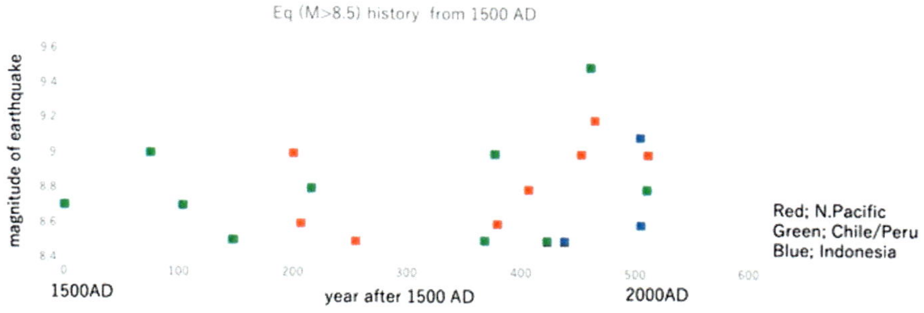

Fig. 3.10 Historical pattern of the giant plate boundary earthquakes surrounding the Pacific Ocean from 1500 to 2018 AD showing the clustering around 1600–1700 and 1900–2018 in the northern Pacific Ocean and Chile/Peru region. Data are cited from JMA data center

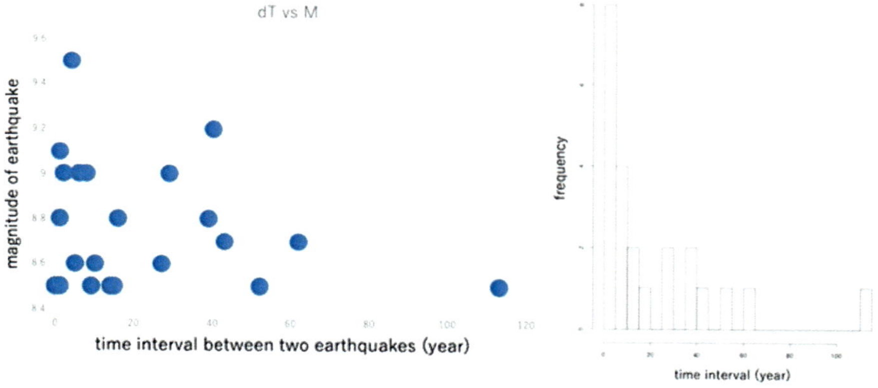

Fig. 3.11 Diagram of magnitude and time interval of the giant earthquakes (left) and frequency of the time interval (right), showing cluster-like events in the short time interval

3.6 Stress Orientation and Seismic Anisotropy of the Plate Boundary

There are three types of stress orientations in the plate boundaries: One is the subduction boundary showing the maximum stress is perpendicular to the trench axis and the minimum is vertical, the second is the oceanic ridge showing the normal fault type stress configuration. The third is the strike-slip type stress configuration is satisfied along the transform fault as manifested by plate tectonics. However, there are many signatures showing normal fault type and strike-slip types earthquakes in the fore-arc regions after giant plate boundary earthquakes such as 3.11, 2011 of Tohoku-Oki earthquake. In addition, the large earthquakes of normal fault type occurred within the subducting Pacific plate at 1973 named as Sanriku-Oki earthquake of which fracture plane cut the plate itself by large normal fault.

On the other hand, there are many normal faults in the fore-arc regions along the Izu-bonin–Mariana trench. Along the fore-arc region near the IBM trench axis, Lau Basin develops between the trench axis and the Izu-Ogasawa island arc suggesting the regional extension basin along the trench axis. Therefore, during the inter-seismic periods, wide regions along the arc and trench axes were suffered from the extensional stress conditions.

Seismic anisotropy may indicate the stress configuration and/or the kinematic axes of the solid-state flow of wedge mantle. In the case of sealed cracks occupied by serpentine and other rock forming minerals such as albite, jadeite, and calcite in the wedge mantle, the anisotropy of S-wave of earthquakes indicates directly the stress configuration such that the slowest orientation of S-wave means the perpendicular direction of sealed crack surface and the fast one involves the parallel orientation of the crack surface. It is because the sealed crack is the type I open crack and it is occupied by sealing minerals. In the case of wedge mantle, the sealing minerals are mainly antigorite serpentine but rarely albite in the shallower part and jadeite in the deeper part.

References

Aki K (1979) Characterization of barriers on an earthquake fault. J Geophys Res 84:6140–6148

Aki K (1984) Asperities, barriers, characteristic earthquakes and strong motion prediction. J Geophys Res 89:5867–5872

Avrami M (1941) Kinetics of phase change. III. Granulation, phase change, and microstructure. J Chem Phys 9(2):177–184

Dziewonski AM, Chou TA, Woodhouse JH (1981) Determination of earthquake source parameters from waveform data for studies of global and regional seismicity. J Geophys Res 86:2825–2853

Ekstrom G (2007) Global seismicity: results from systematic waveform analyses, 1976–2005. In: Kanamori H (ed) Treatise on geophysics. Elsevier, pp 473–481

Friedlander SK (1977) Smoke, dust, and haze: fundamentals of aerosol behavior. Wiley, New York, p 400

Fujie G, Kodaira S, Sato T, Takahashi T (2016) Along-trench variations in the seismic structure of the incoming Pacific plate at the outer rise of the northern Japan trench. Geophy Res Lett 2016(43):666–673

Green II HW, Burnley PC (1989) A new self-organizing mechanism for deep-focus earthquakes. Nature 341(6244):733–737

Gutenberg B, Richter CF (1944) Frequency of earthquakes in California. Bull Seismol Soc Am 34:185–188

Ide S, Shelly DR (2007) A scaling law for slow earthquakes. Nature 447:76079. https://doi.org/10.1038/nature05780

Igarashi T, Matsuzawa T, Hasegawa A (2003) Repeating earthquakes and interplate aseismic slip in the northeastern Japan subduction zone. J Geophys Res 108(B5):2249. https://doi.org/10.1029/2002JB001920

Ito E, Sato H (1992) Effect of phase transformations on the dynamics of descending slab, In: Syono Y, Manghnani MH (eds) High pressure research: application to earth and planetary sciences, pp 257–262

Ito Y, Obara K, Shiomi K, Sekine S, Hirose H (2007) Slow earthquakes coincident with episodic tremors and slow slip events. Science 315:503. https://doi.org/10.1126/scienc.1134454

Kanamori H (1986) Rupture process of subduction zone earthquakes. Ann Rev Earth Planet Sci 14293–14322
Kanamori H, Anderson DL (1975) Theoretical basis of some empirical relations in seismology. Bull Seismol Soc Am 65:1073–1095
Kanamori H, Stewart GS (1978) Seismological aspects of the Guatemala earthquake of February 4. J Geophys Res 83:3427–3434
Karato S (2008) Deformation of earth materials: an introduction to the rheology of solid earth. Cambridge University Press, p 463
Kato A, Obara K, Igarashi T, Tsuruoka H, Nakagawa S, Hirata N (2012) Propagation of slow slip leading up to the 2011 Mw 9.0 Tohoku-Oki earthquake. Science 335:705–708. https://doi.org/10.1126/science.1215141
Ogata Y (1988) Statistical models for earthquake occurrences and residual analysis for point processes. J Am Stat Assoc 83:9–27
Ohnaka M (2000) A physical scaling relation between the size of an earthquake and its nucleation zone size. Pure Appl Geophys 157:2259–2282
Ohnaka M, Matsu'ura M (2002) The physics of earthquake generation. Univ. Tokyo. Press, p 378. in Japanese
Oliver J, Isacks B (1967) Deep earthquake zones, anomalous structures in the upper mantle, and the lithosphere. J Geophys Res 72:4259
Omori S, Kamiya S, Maruyama S, Zhao D (2002) Morphology of the intraslab seismic zone and devolatilization phase equilibria of the subducting slab peridotite. Bull Earthq Res Ints Univ Tokyo 76:455–478
Ringwood AE (1975) Composition and structure of the earth's mantle. McGraw-Hill
Schwartz S, Rokosky JM (2006) Slow slip events and seismic tremor at circum-Pacific subduction zones. Rev Geophys 45:RG3004. https://doi.org/10.1029/2006rg000208
Toriumi M (1986) Mechanical segregation of garnet in synmetamorphic flow of pelitic schists. J Petrol 27:1395–1408
Toriumi M (2008) A viewpoint of nonlinear dynamics to earthquakes. Kagaku (Science in Japan) 78(11):1233–1237 (in Japanese)
Utsu T (1961) A statistical study on the occurrence of aftershocks. Geophys Mag 30:521–605

Chapter 4
Data-Driven Science for Geosciences

Abstract Recent development of data-driven sciences gives a wide applicability for basic and applied sciences for natural phenomena with and without known basic models. Multivariate time series and image data with huge amounts of freedom are now investigated by means of decomposition methods of data matrix and state-space modeling to infer macroscopic invariances in the natural phenomena using sparse modeling.

Keywords Matrix decomposition method · Sparse modeling with LASSO · State-space modeling of time series

4.1 Matrix Decomposition Method and Sparse Modeling

Natural sciences concerned with natural substances and phenomena are often divided into descriptive science and model-oriented science: The former is applied for the complicated and time-dependent natural systems such as biological phenomena, and the other is for the simple physical phenomena such as geosciences and planetary sciences having something basic to be hidden universalities or equations. Many scientists have investigated the experiments analogous to natural phenomena in the universe; however, the geological and astronomical phenomena and substances contain huge amounts of elementary processes and huge ranges of time scale variations controlling such phenomena. Furthermore, it is very difficult to construct the available and universal simple models because natural phenomena do not repeat the same processes, and they involve the mutual correlated processes and parameters in the history from early to late universe. On the other hand, we can conduct the rigorously repeatable experiments for the artificial system in the small systems and high energy systems within the very short time scales. Exceptional cases of

Electronic supplementary material The online version of this chapter (https://doi.org/10.1007/978-981-15-5109-3_4) contains supplementary material, which is available to authorized users.

M. Toriumi, *Global Seismicity Dynamics and Data-Driven Science*, Advances in Geological Science,
https://doi.org/10.1007/978-981-15-5109-3_4

the natural phenomena are the astronomical and mechanical processes of planet and satellite motion in the solar system which are almost always governed by the classical mechanics.

Solid earth phenomena involve many elementary mechanics such as fluid dynamics, solid mechanics, thermodynamics, reaction kinetics, electromagnetic fluid dynamics, solid-state physics, materials sciences, biosciences, isotope geochemistry, high-pressure physics, and others. These disciplinary basics involve also the nonlinear kinetics resulting the chaotic behaviors as the phenomena. Therefore, it is suggested that to understand the natural phenomena and to control and predict the processes in the human system, it is needed to make the model available for the artificial systems in the human community by means of mathematical formulations under combination of the disciplinary elementary processes such as stochastic prediction of large earthquakes associated with huge disaster.

Models available for taking benefits in the human society are better in the minimum parameters than in the more complicated parameters as mentioned in the Occum's Razer, but in the natural systems, we need the complicated models for the mimics of the natural processes because of complicated nonlinear behaviors of the natural systems. Thus, it seems that the available models should contain the many parameters and nonlinear terms in the system equations, and thus, they need huge amounts of data to determine the effective parameters by data assimilation.

Data assimilations are general method to obtain the actual equations available to apply to the natural system such as weather prediction using aerodynamics. In the continuum mechanics involving phase transition and chemical reactions, the modeling of the inhomogeneous systems is very complicated because of mathematically infinite dimensional system and contains the data assimilation of vast amounts of parameters. For examples, we may imagine the tomographic study of the solid earth interiors by means of seismic wave velocity change. Then, the best approximated parameters set of the earth interiors are imaged in the 3D mapping (Fukao et al. 1992). In this method, we should be required for vast amounts of seismic data obtained by huge global networks of seismic stations distributed in the world introduced in the previous chapter. Recent data assimilation methods for these tomographic studies are based on the Bayesian inversion technique with Markov chain Monte Carlo (MCMC) calculation to search the global minimum conditions of parameters set as proposed by Kuwatani et al. (2018).

Data-driven sciences involving Bayesian inversion technique with MCMC are widely applied to the natural and technological sciences, and in these newly developing sciences, the sparse modeling and dimension reduction methods for huge amounts of data sets are very available for formulating the mathematical modeling of complicated apparent high dimension systems as noted by Bishop (2006). Classically speaking, the dimension reduction methods as one of sparse modeling are classified into principal component analysis (PCA) and independent component analysis (ICA), and the former is based on the eigenvectors with large eigenvalues determined by symmetric covariance matrix of huge data sets assuming the Gaussian distribution function, but the latter on the best fit assuming exponential type distribution function by the oblique coordinate basis. For examples, Iwamori and Nakamura (2015) succeeded in classifying the source mantle processes controlling the multi-components trends of primitive magmas from the mid-oceanic and oceanic island basalts: The one is derived from wet deep mantle, and the other is from relative dry mantle source. It is very important that the spatial distribution of the former type is limited in the deep mantle from western half from the central Pacific Ocean (180 E) and that of the latter is another eastern half deep mantle of the solid earth. This means that the structure of the geochemical signals is forming first-order longitudinal inhomogeneity of the deep mantle. It is probably the inhomogeneous stagnant distribution of subducted slabs in the deep mantle until now which should contain large amounts of hydrated mantle materials. Furthermore, recent investigations of network GPS using ICA analyses of whole China revealed several modes of regional-scale periodic deformation (vertical and horizontal displacement) by Ming et al. (2017). They obtained the results that the spatial characteristic displacement patterns show NS and EW trends by means of ICA instead of PCA assuming non-Gaussian-type signals of GPS.

On the other hand, the author tried to investigate the PCA filtering of seismicity density distribution patterns in the SW Japan and NE Japan regions, and obtained the results that the long-term trends of the major components and annual variations of the higher-order components (Toriumi 2009, 2011; Okada et al. 2017). These difference between ICA and PCA is derived probably from distribution density function of the data: The former is available for the case of non-Gaussian-type but the latter is for the case of Gaussian (lognormal) type.

Sparse modeling method is developed by Nakamura et al. (2015) in the field of the episodic sedimentation processes such as tsunami deposits after 3.11 of 2011 Tohoku-Oki giant earthquake by means of geochemical multi-components clustering and by Ueki and Iwamori (2017) for the geochemical classifications of whole crustal rocks. They clarified that the several characteristic chemical compositions are effective in classification of the tsunami sediments and others and obtained the spatial distribution of tsunami sediments in the on-land regions apart from the coast line.

Sparse modeling is developed by stochastic PCA method with least absolute shrinkage and selection operator (LASSO) for multi-components classification of various type data sets. This method is based on the Lagrange function of L2 norm of data and transferred data + L1 norm of base vectors to be minimum, so that the several effective components of original base vectors can be significant to modeling as introduced in the later chapters. Recent development of data processing without physical model appears the non-negative matrix factorization method (called as NMF method) in which data matrix is to be decomposed into coefficient matrix and basis matrix under their non-negative components. This method is based on the basis matrix

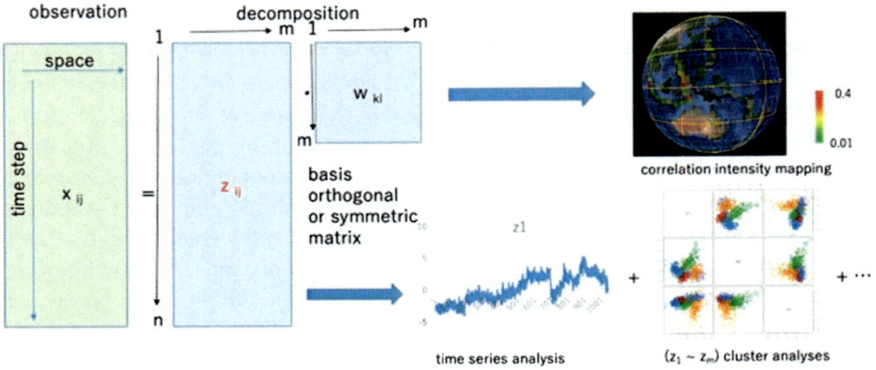

Fig. 4.1 Illustration of the procedures of observation data through data-driven analytics using the decomposition by basis orthogonal (PCA) and symmetrical (non-negative factorization; NNF, independent component analysis; ICA) matrix

expression in the oblique coordinate system (that is not orthogonal), and thus, NMF is different in taking the basis matrix at the decomposition but resembles with the method of ICA.

In summary, the recent methods of the machine leaning for reconstructing the huge amount of noisy observation data are fundamentally based on the decomposition and spectrum decomposition of data matrix into the basis matrixes and their coefficient matrixes as shown in Fig. 4.1. The case of orthogonal basis matrixes of decomposed matrix is equivalent to the SVD and PCA, but the case of oblique basis matrixes is the NMF and ICA. The regularization techniques are recently added in the optimization for minimum Lagrange function (cost function) with L1 (LASSO) and L2 norm conditions and some variety. In many cases, the data matrix decomposition is used to be linear algebra but in the case containing nonlinear transformation the deep neural network decomposition with many intermediate layers and nodes may be favorable for the detailed study of the huge data system together with state—space modeling method (Commandeur and Koopman 2007; Kitagawa 2010).

4.2 Deep Neural Network Approximation

In order to transform the huge amounts of data to important and robust correlated variables involving the invariant relations, we may refer the encoder using several linear and nonlinear functions: ICA and PCA are the method of linear transformation under the maximum dispersion of data matrix in the case of non-Gaussian and Gaussian fluctuations of data, and neural network deep machine learning methods (artificial intelligence; AI) are the nonlinear transformation one. As is well known, the neural network is composed of many layers of multiple nodes of signal input and output and these layers are mutually connected by the nonlinear transfer function such as hyperbolic tangent type. Thus, every node in the layer receive the input signals from the nodes of the former layers and send the output signals to nodes of the next layer. The deep neural network model involves large number of layers and thus huge amount of parameters to be learned in the data matrix. The DNN encoder is constituted with the data input layer, medium layers and output layer.

If we make the system connecting encoder and decoder, that is the reverse of encoder, the system is called as the recurrent neural network (RNN) and it generates the transformed input data. Then, it is possible to compare the difference between them in the case of given RNN parameters which control the nonlinear transformation (activation) functions between each node. In the case of many medium layers in DNN, the most available parameter sets should be inferred by taking minimum value of the following Lagrange function (Bishop 2006) as

$$L = \sum_{(i,j)} (x - \mathrm{RNN}(x))^2 \tag{4.1}$$

In the equation, RNN(x) is the decoded value from initial input data x. The huge amounts of parameters in activation function from node to node of layer to layer should be determined by huge amounts of input data x. If the input data are selected as subsets of whole data, we can obtain the different subsets of large amounts of activation parameters to store the learning memories in the neural network. This is called as the convolution neural network (CNN) and is used to image reconstruction and coloring.

Now, if we consider the multiple time series data, the input data of the time ($t - m$; $m = 0, 1, 2\ldots$) may be transformed into the output data of the time $t + 1$ (Fig. 4.2). The machine learning of huge amounts of time step samples of observed

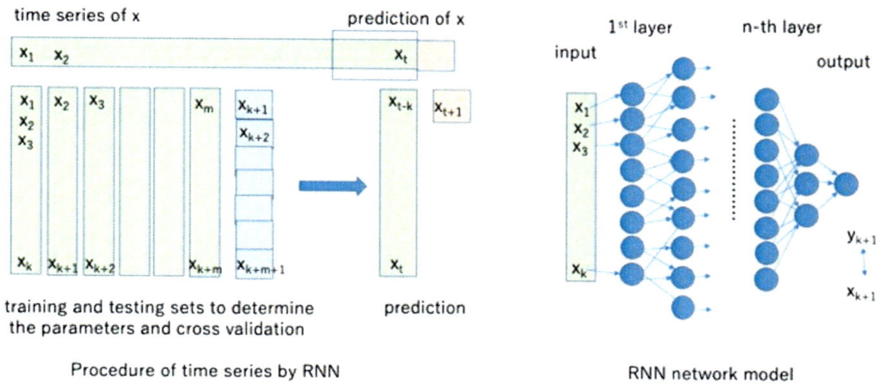

Fig. 4.2 Illustration of the deep learning recurrent neural network (RNN) method for prediction of time series, showing the divided sets of observed time series into training set, testing set of partial segments of time series, and the prediction set. Right: illustration of input and output with layer structure involving nodes connecting layers and of single output y is to be compared with x

data should be used to learn the huge amounts of activation parameters in DRNN, and furthermore, the near-future prediction of the time series can be performed in this DRNN with machine-learned parameters (see Appendix I).

4.3 State-Space Modeling of Time Series

The time series of the huge amounts of data is expressed by big data matrix having huge rows and columns in which the time steps are row vectors with column components. This type of multiple time series is possibly treated by the method of state-space modeling (SSM) with Kalman smoothing and filtering (Kitagawa 2010). In the SSM modeling of time series, the data vectors of the time t is expressed by the internal state variable of the same time with effective noise and the internal state variable at the time t is controlled by that at the time $t - 1$ with some transition probability w and actual noise. The following equations should be satisfied as like as

$$
\begin{aligned}
x(t+1) &= H_t x(t) + p(t) \\
y(t) &= W_t x(t) + q(t)
\end{aligned}
\tag{4.2}
$$

In which H and W are the time dependent transformation matrixes, respectively, and p and q are noise terms associated with state space and observables, respectively. In the form of discrete variables, above equations are rewritten as

$$
\begin{aligned}
x_i &= H\, x_{i-1} + p_i \\
y_i &= W\, x_i + q_i
\end{aligned}
\tag{4.3}
$$

As seen in these equations, H is the transition probability from $(i - 1)$-th to i -th states as like as Markov chain, and W means the transformation matrix from hidden states to observable y (Fig. 4.3). We concern the time series of the multi-components system describing the seismicity of the solid earth. Considering the global seismicity, the hidden state variables may be taken to be the mechanical states such as stresses

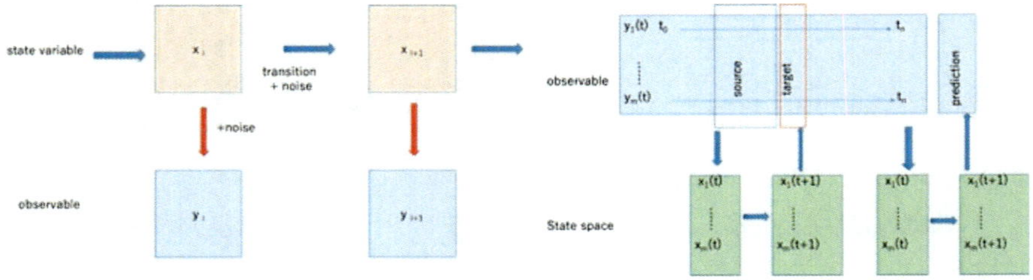

Fig. 4.3 Left: Illustration of state-space modeling (SSM) method for smoothing and prediction of the time series showing the transition process in the hidden state and transformation to time series of observed variables. Right: actual procedure of SSM modeling in the global seismicity dynamics in this book

and strains active in the rock masses. These mechanical parameters responsible for the seismic faults or shear cracks should vary along with the plate boundary zones and also intraplates although there is an intrinsic mechanical interaction among various localities of these regions because of the mechanical continuity of the solid earth.

The SSM modeling of multiple time series is available for the prediction of the global seismicity trends in near future by means of the explicit representation of the time series instead of the DRNN method because the DRNN analyses are used to take a black-box-type transformation of data obtained by deep learning. On the other hand, the SSM is based on the obvious hidden state change with time-dependent Gaussian noises as shown in Fig. 4.3. Later, the author introduced the procedures of application for SSM to investigate the possibility of stochastic prediction of the global and regional seismicity.

The SSM method to obtain the complicated trends of the high-dimensional seismicity time series is required to combine the dimension reduction processing of the high dimension data by means of PCA or the sparse modeling using LASSO. In these cases, the several principal components by linear combination of original high dimension data should be taken as the lower dimensional time series to be processed to SSM modeling. Furthermore, as the stochastic PCA and sparse PCA are both linear transformation by the reduced eigenvectors matrix of covariant matrix from data matrix, the inverse matrix of transformation matrix is easy to obtain, and thus, the stochastics seismicity prediction of each original components can be inferred by the inverse matrix as follows;

$$x_j(t) = W^{-1} z_j(t) + W^{-1}\,\text{noise} \tag{4.4}$$

In this equation, the inverse matrix W^{-1} which is equal to W^T because of orthogonal matrix has abundant very small components, and z_j with j more than 5 is nearly Gaussian noise having mean value = 0. So, the noises of each local seismicity have a mean value of zero but the variance which is equal to the sum of the variances of z_j variation. If every variances of the z_j are equal to σ, we can obtain the total variance of x_j about σ because of total W^{-1} components equal to unity. If the variances of each z components are different, the variance of each local seismicity x_j is the weighted average of σ_j.

4.4 Frobenius Norm Minimum Method for Dynamics

Let us consider the system dynamics having huge data matrix $\{x_{ij}\}$ for i = time and j is locality index. In the studied system, the locality index must be taken as the number of neighboring ones because of the gradient controlling system dynamics. This is the simplified method for searching a macroscopic equation in the data matrix proposed by Schmidt and Lipson (2009) and Brunton Proctor and Kutz (2016).

Then, we possibly assume the following dynamic equation in the matrix form;

$$\{x_{i,j}\} + \alpha\left[\{x_{i+1,j}\} - \{x_{i,j}\}\right] + \beta\left[\{x_{i,j+1}\} - \{x_{i,j}\}\right] = N(0, \sigma^2) \tag{4.5}$$

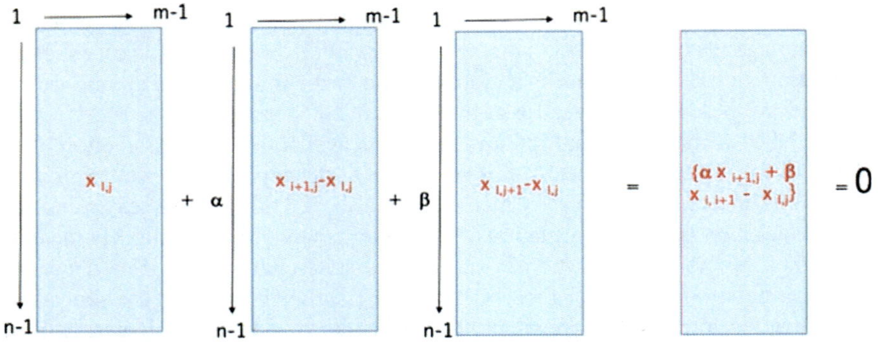

Fig. 4.4 Illustration of Frobenius norm minimum method for modeling the dynamical equation from the spatio-temporal variation of observed data matrix

In this equation, the second term means the time derivative, and the third term does the space derivative. N (0, σ^2) means the Gaussian noise with mean $= 0$ and variance $= \sigma^2$. Excluding the constant term gamma, the above linear combination of three matrixes should become zero in the available model (Fig. 4.4). Then, the following evaluation function should be minimum on the parameters of α and β in the condition of Frobenius norm minimum as

$$J = \sum_{(i,j)} (x_{ij} + \alpha' x_{i+1,j} + \beta' x_{i,j+1})^2$$
$$dJ/d\alpha' = 0 \text{ and}$$
$$dJ/d\beta' = 0 \tag{4.6}$$

where α' and β' are $\alpha/(1 + \alpha + \beta)$ and $\beta/(1 + \alpha + \beta)$, respectively.

However, as far as the data have huge amounts of components, it is quite difficult to obtain the minimum value of the above Frobenius norm as shown in Fig. 4.4. Therefore, preprocessing the primary data by dimension reduction using deterministic PCA or sparse PCA, it is easy to obtain the minimum conditions for Frobenius parameters α and β. In this book, the author intends to infer the minimum dynamics by means of simple data assimilation using stable nodes of the attractors in the reduced parameter space because of nonlinearity of the global seismicity mechanics.

It is noted that the NMF method described previously is also based on the Lagrangian form of cost function between data matrix D and basis matrix decomposition WU as well as dynamic mode decomposition (DMD) method (Durbin and Koopman 2012).

References

Bishop CM (2006) Pattern recognition and machine learning. Springer, New York

Brunton SL, Proctor JL, Kutz JN (2016) Discovering governing equations from data by sparse identification of nonlinear dynamical systems. PNAS 113(15):3932–3937

Commandeur JJF, Koopman SJ (2007) An introduction to state space time series analysis. Oxford University Press

Durbin J, Koopman SJ (2012) Time series analysis by state space methods. Oxford University Press

Fukao Y, Obayashi M, Inoue H, Nenbai M (1992) Subducting slabs stagnant in the mantle transition zone. J Geophys Res 97(B4):4809

Iwamori H, Nakamura H (2015) Isotopic heterogeneity of oceanic, arc and continental basalts and its implications for mantle dynamics. Gondwana Res 27(3):1131–1152

Kitagawa G (2010) Introduction to time series modeling, monographs on statistics and applied probability 114, CRC Press, 289 pp

Kuwatani T, Nagata K, Yoshida K, Okada M, Toriumi M (2018) Bayesian probabilistic reconstruction of metamorphic P-T paths using inclusion geothermobarometry. J Mineral Petrol Sci 113:82–95. https://doi.org/10.2465/jmps.170923

Ming F, Yang Y, Zeng A, Zhao B (2017) Spatiotemoral filtering for regional GPS network in China using independent component analysis. J Geodesy 91(4):419–440

Nakamura K, Kuwatani T, Kawabe Y, Komai T (2015) Extraction of heavy metals characteristics of the 2011 Tohoku tsunami deposits using multiple classification analysis. Chemosphere 144:1241–1248

Okada A, Toriumi M, Kaneda T (2017) Spatial and temporal pattern of global seismicity extracted by dimensionality reduction. Int J Geology 11:26–34

Schmidt M, Lipson H (2009) Distilling free-form natural laws from experimental data. Science 324(3):81–85. https://doi.org/10.1126/science.1165893

Toriumi M (2009) Principal component analyses of seismic activity in the plate boundary zone of northeast Japan arc. J Disaster Res 4(2):209–213

Toriumi M (2011) Dimension reduction study of microseismic activity in the earth's crust and mantle in the plate boundary region, In: Multiscale mathematics: hierarchy of collective phenomena and interrelations between hierarchical structures, COE Lecture Note vol 39, of Kyushu Univ., pp 116–131

Ueki K, Iwamori H (2017) Geochemical differentiation processes for arc magma of the Sengan volcanic cluster, Northeastern Japan, constrained from principal component analysis. Lithos 290–291:60–75

Chapter 5
Data-Driven Science of Global Seismicity

Abstract It is well known that the seismic activity is basically constraint by the plate tectonics framework and that the elastic interaction of the plate in the global earth is transmitted within 10^4 s. Thus, the global seismic activities of every plates and plate boundaries may be strongly correlated with each other. Those correlation patterns may be a kind of invariances in the dynamics of the global seismicity.

Keywords Data cloud of seismicity · Correlated seismicity of logarithmic seismicity rate · Characteristic features of global correlated seismicity rates

5.1 Data Cloud of the Global and Japanese Seismicity

First, we intend to define the seismicity rate by the number density of the earthquake source events in the given volume and time duration in the solid earth. Here, the earthquake events are screened by their magnitude, for example, that is the magnitude range of M4–M5, M5–M6, and so on. The unit volume of counting earthquake source events is selected as the width of 2.5° of latitude and 5° of longitude, and depth range from the earth surface to 700 km depth along the subduction zones and mid-oceanic ridges in the world as indicated in Fig. 5.1. It is the reason to select the size of seismicity volume here why the stochastic investigation of the seismicity rate is required for large source events number of each volume and for large number of studied volume dividing the earth crust and upper mantle in order to increase the spatial resolution. To increase the time resolution is also needed to investigate the detailed change of seismicity rates near the large earthquakes, but it should be avoided for the small number of seismic events counted in. So the author investigates time ranges of the seismicity rate to be several cases such as 3, 10, 15, 20, 30, and 50 days

---

Electronic supplementary material The online version of this chapter (https://doi.org/10.1007/978-981-15-5109-3_5) contains supplementary material, which is available to authorized users.

M. Toriumi, *Global Seismicity Dynamics and Data-Driven Science*, Advances in Geological Science,
https://doi.org/10.1007/978-981-15-5109-3_5

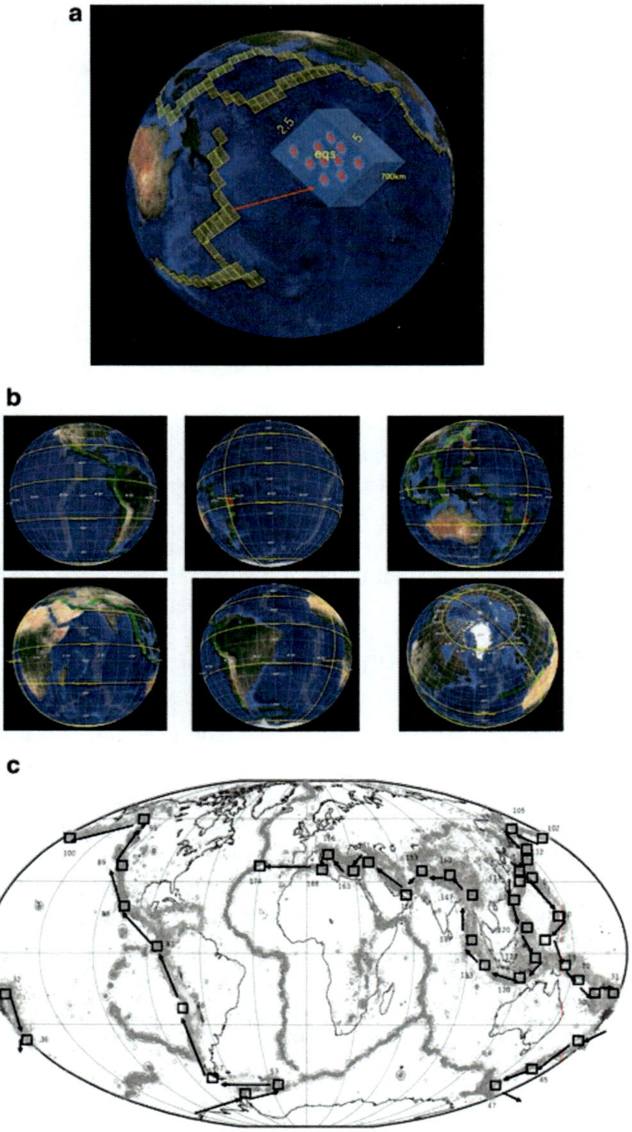
a
2.5
5
eqs
700km
b
c

◄**Fig. 5.1** **a** Illustration showing the locality cell in the global subduction and oceanic ridge zone for measurement of seismic event density defined as event number per unit time and area of longitude 5° and latitude of 2.5°. **b** Positions of locality cells in the global subduction shown by green to red quadrangle and oceanic ridge zones by white quadrangle in the Google Earth. **c** Position numbers of locality cells along the global subduction zones, showing the quadrangles with numbers from central Japan to New Zealand, Chile, Alaska, and again to NE Japan, and through Indonesia, Himalaya, and to northern Atlantic Ocean from 1 to 174. **d** Position number is shown by yellow of locality cells along the global oceanic ridge zones from North Pole region to Iceland, central Atlantic Ocean, South of Cape Town, central Indian Ocean, Red Sea, and again central Indian Ocean, South of Tasmania, South of New Zealand, West of Chile, and to Off Mexico

in order to compare the Bayesian information criteria (BIC; Bishop 2006) of the cluster analyses of the seismicity data.

The huge amounts of seismicity rates are obtained by sort processing of databases of USGS, ERI, and NIED database centers in the R-language. The data structure in these database is the time successive list in the order of the event clock of year/month/day/hour/minute/second and hypocenter coordinates (longitude, latitude), depth, magnitude, and their errors and comments. In this book, the data are used from January 1, 1990 to June 30, 2018 in the global earth and in the regional cases of Japanese regions these are used from January 1, 1998 to September 29, 2018 because of homogeneous data errors as noted in the previous sections. The time series of the data is used as the number of days after January 1, 1990 because of the convenience for counting the clocks.

In this study, the global seismic data stored in the database of ISC in ERI seismicity data center and USGS data catalog are used for uniform data error for earthquake hypocenters determining by CMT method. Huge amounts of seismic source data in the database are sorted in order to make data matrixes of event numbers of earthquakes ranging from magnitude 4–5 and 5–6 and 7.5–9.5 in the unit volumes (called as the locality cell in this book) of 2.5° of latitude, 5° of longitude, and depth of 700 km taken along the subduction zones and mid-oceanic ridge zones in the world, because hypocenters of earthquakes determined by the global seismic network under magnitude of 4 are not available distribution in the subduction zones apart from dense network stations in the world. The localities data sets of the subduction zones and the mid-oceanic ridge zones are listed in the locality cell directory of the Electronic Supplement Materials (ESM) and are shown on the world map as shown in Fig. 5.1.

In the new database sets, seismic events tagged by year, month, latitude, longitude, depth, and magnitude are prepared for data-driven scientific researches in order to investigate the global seismicity dynamics. The new database obtained here is composed of subsets of database: Pacific subduction zone, Philippine Sea, Indian Ocean, Mediterranean Sea subduction zones, Mid-Atlantic Ridge, Indian Ocean Ridge, and East Pacific Rise regions in the world. Subsets of global events of the large earthquakes ranging from magnitude 7.5 to 9.5 are also made in this study. These time series of seismicity in subsets of database in the directories of ESM are shown in Fig. 5.2.

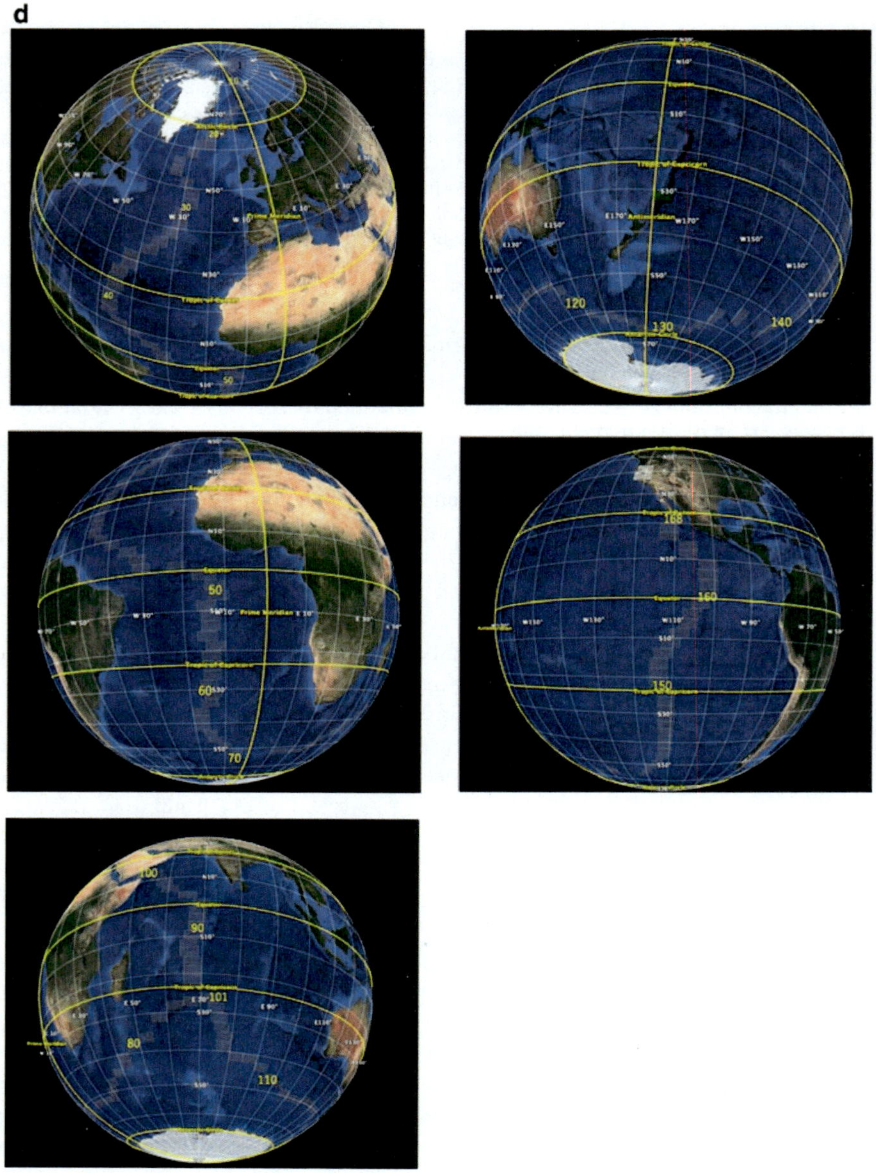

Fig. 5.1 (continued)

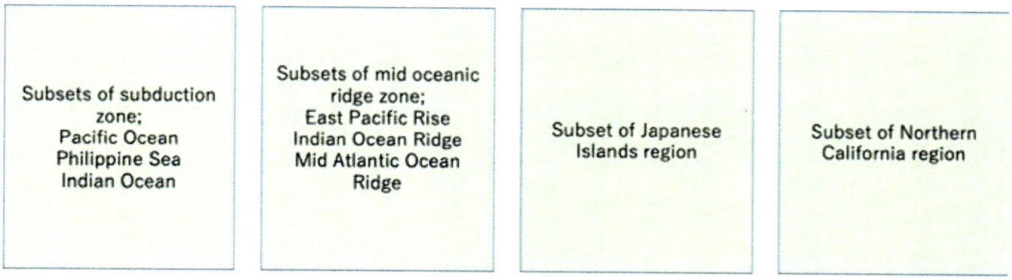

Fig. 5.2 Illustration of data sets of the seismic event density time series in the global and regional seismic active zones

In the regional seismicity database in the Japanese islands region, seismicity event numbers in the cells are sorted by the method described above from the primary database stored in JMA-1 in ERI and database of National Institute of Earthquake Disasters (NIED). The magnitudes of the sorted earthquakes are ranged from 1.0 to 2.0, 2.0 to 3.0, and 3.0 to 4.0 and from 1998/1/1 to 2018/9/29 because of network foundation of digital seismic stations at 1998. The seismicity under magnitudes of 2.0 is usually thought to involve some amounts of observation loss noted previously, but the seismicity rate data used in this book have precise event clocks and comparative large locality cell volumes covering any amount of hypocenter noises. Furthermore, the author tried to analysis by means of data-driven scientific methods for data of logarithmic number density of earthquakes and thus the loss term coefficient should be deleted by the difference from the mean value.

The time series of the seismicity rates for every cell studied here are shown in Fig. 5.3, indicating the apparent large variations of the patterns of the time sequences. To avoid the complicated time series of large seismicity and small one in the various locality cells, several representative time series of raw seismicity rates were displayed in Fig. 5.4. It seems that there are several patterns of time sequences of these seismicity rate changes in the subduction zones of the world. The one shows the high and continuous activity appeared in the NE Japan, Aleutians, and Java-Sumatra. The weak activity appears in the SW Japan, Izu-Ogasawara, Kermadic–New Zealand, and Middle America. The others show the intermediate activity. The seismicity rates in the mid-oceanic ridge regions are relatively weak as shown in Fig. 5.5, likely to the seismicity rates of SW Japan and others.

Combining the seismicity rates of various locality cells at the time t, we define the global seismicity rate vector $X(t)$ as

$$X(t) = \{x_j(t)\} = \{\log(1 + n_j(t))\} \tag{5.1}$$

where n_j and x_j are the number of seismic events and natural logarithmic seismicity rate in the jth locality cell, respectively, and t is the time steps from January 1990 to December 2018. The time intervals are taken as 1 day, 3 days, 5 days, 10 days, 15 days, 30 days, and 50 days, and in the later chapters, the author tried to check the validity of the time intervals available for the data-driven scientific studies as

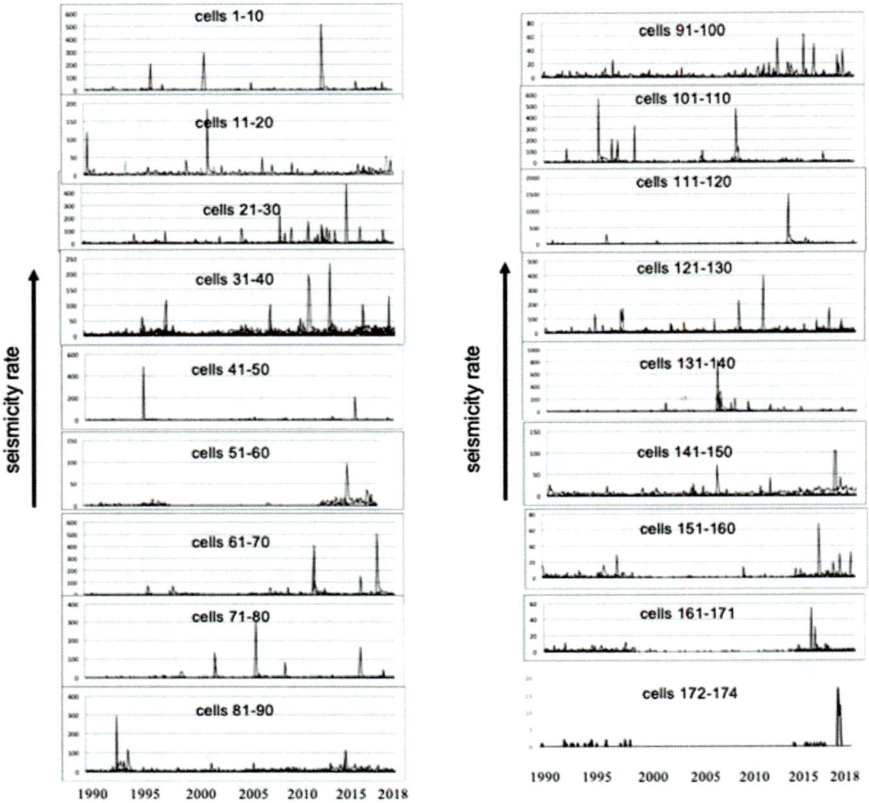

Fig. 5.3 Time series of the seismic event density in all locality cells along the global subduction zones from January 1, 1990 to June 25, 2018. Each diagram contains the time series of 10 locality cells

illustrated in Figs. 5.6 and 5.7. As a result, the data sets of the 10 days interval are most available for decomposition methods of data matrixes studies by deterministic PCA and stochastic PCA with LASSO because of huge size of data (Fig. 5.8).

As stated previously, the observation loss in the number of micro-earthquakes under magnitude 2 affects the apparent number of them should be given by

$$n_i(t) + 1 = \phi(1 + n_{i,\mathrm{ap}}(t)) \tag{5.2}$$

where $n_{i,\mathrm{ap}}$ is the apparent number of earthquakes in the range of M1–M2, and ϕ is the coefficient of observation loss about 1.13 in the case of Southern California (Christensen et al. 2002) and 1.2 in the case of Japanese islands region derived from the seismic network system which is probably independent on time and space (Fig. 5.9). Then, we obtain the following logarithmic seismicity rates of earthquakes between M1 and M2 in the case of Japanese islands region,

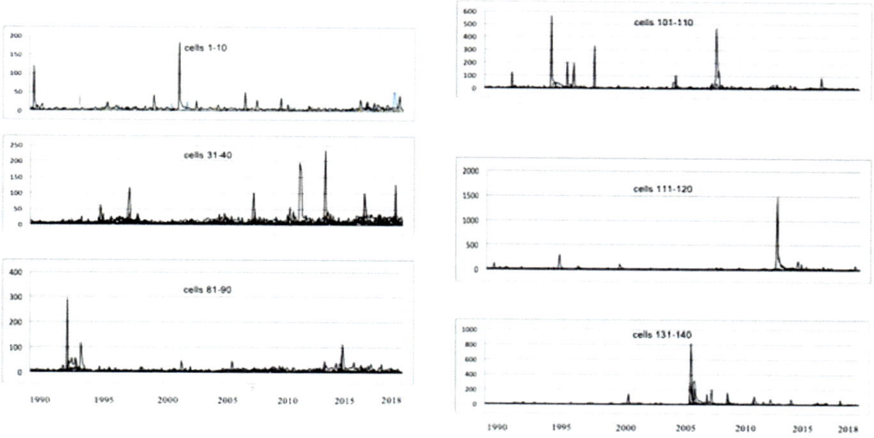

Fig. 5.4 Time series of the seismic event density in all locality cells along the global oceanic ridge zones from January 1,1990 to June 25, 2018. Each diagram contains the time series of 10 locality cells

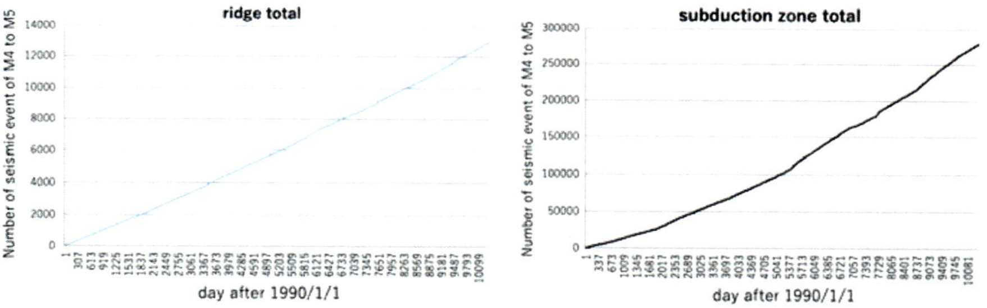

Fig. 5.5 Cumulative total seismic event numbers of M4–M5 in the oceanic ridge and subduction zones

$$x_i(t) = \log\bigl(1 + n_{i,\mathrm{ap}}(t)\bigr) + C \tag{5.3}$$

where C is the constant term related to the observation loss by seismic network system.

In order to investigate the stochastic processes of the global and regional seismicity dynamics, the probability density patterns of the seismicity should be clarified by studying the frequency distributions of the logarithmic seismicity rates defined above. In Fig. 5.10, the frequency distributions of several representative locality cells are shown. It strongly suggests that every frequency distribution patterns of the logarithmic seismicity rates (density) are apparently the same as the Gaussian and multiple Gaussian distribution patterns. It means also that the logarithmic inter-spike intervals of micro to small earthquakes in the locality cells show the Gaussian distribution pattern. Therefore, the time series analyses of the logarithmic seismicity rates

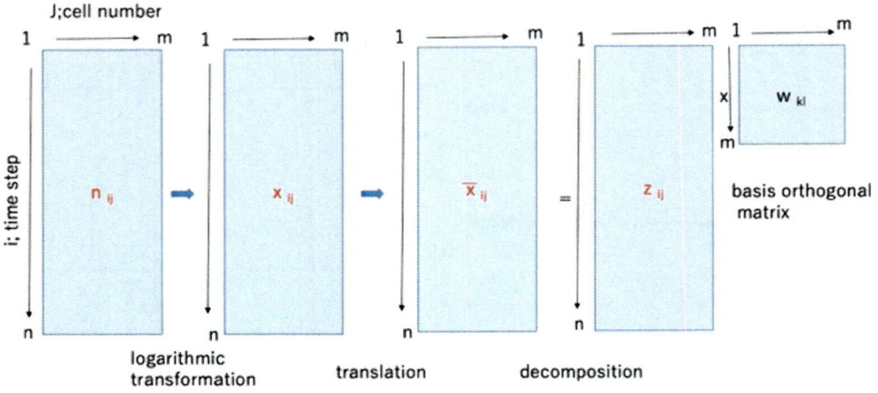

Fig. 5.6 Illustration of procedures in data matrix processing for time series of the global and regional seismic event density, showing three steps of logarithmic transformation, translation by mean value, and decomposition by basis orthogonal matrix

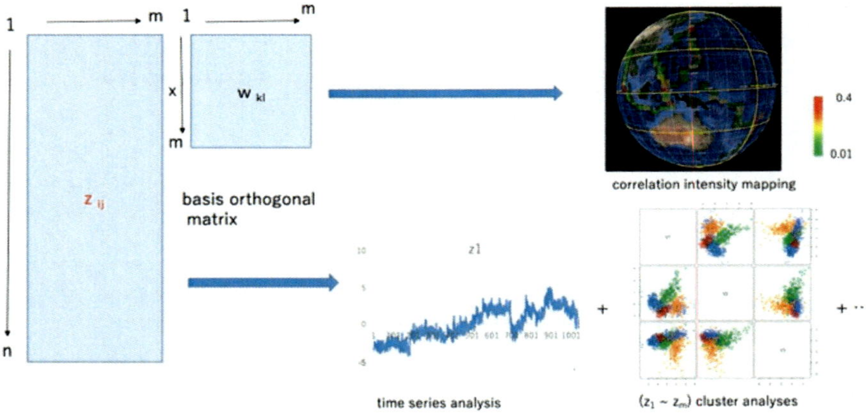

Fig. 5.7 Illustration of data processing of global seismicity rate in the subduction and oceanic ridge zones by decomposition of data matrix into correlated seismicity rates z and basis orthogonal matrix w. The correlated seismicity rates z show time series and basis matrix does correlation locality cells

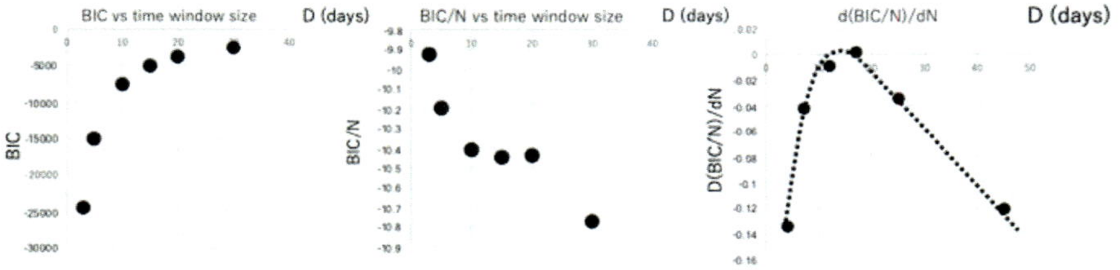

Fig. 5.8 Diagrams of Bayesian information criterion (BIC), time window size (D) of seismic event density, and number of locality cells (N), showing the maximum of the rate of BIC against N

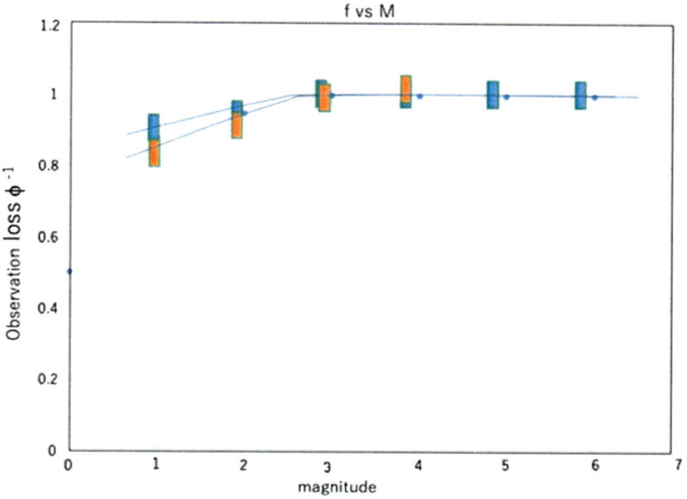

Fig. 5.9 Relations of the observation loss dependent of the seismic station network system and magnitude. The observation loss is defined by the ratio between the extension of GR law to the micro-earthquakes and the observed frequency of them (see text). Blue; Southern California network, orange; Japan network

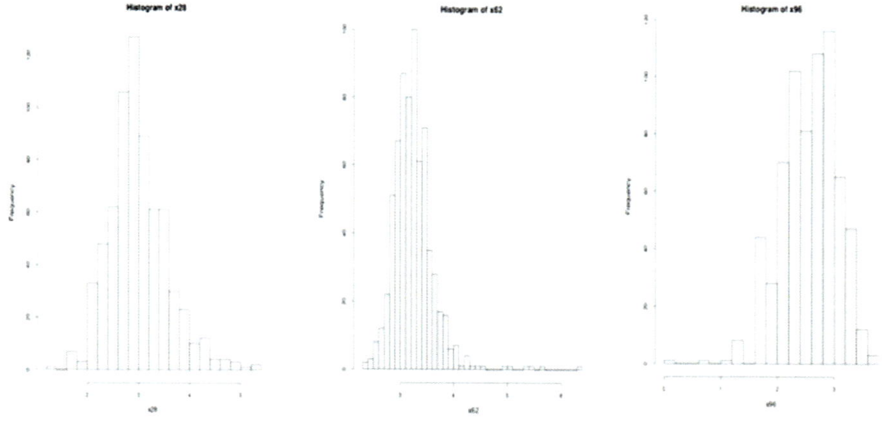

Fig. 5.10 Diagrams of logarithmic time intervals (horizontal axis) and their frequency (vertical axis) between small earthquakes (M1–M2) in various locality cells in the Japanese region, showing the Gaussian-type function

should be available for the reconstruction methods of basis matrix using singular value decomposition (SVD), principal component analysis (PCA), and non-negative matrix factorization (NMF) as illustrated in Fig. 4.1.

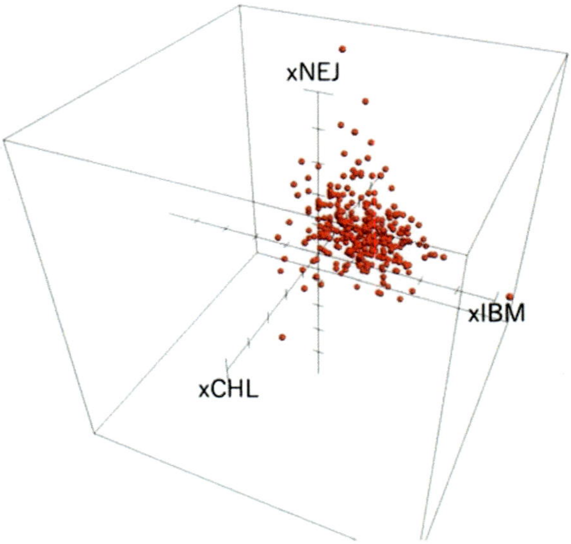

Fig. 5.11 Three dimensions representation as the part of the high dimension logarithmic seismicity cloud in the subduction zones showing the dense core and surrounding scattered data points. NEJ; Northeast Japan, CHL; Chile, IBM; Izu-Bonin-Mariana

The data clouds in this study will be introduced as shown by the data point set of seismicity rates. For visual convenience, we will display the 3D image of the data cloud of the global seismicity case by taking the locality cells of NE Japan, Tonga, and N. Chile in Fig. 5.11. It shows that the central spheroid containing most of data points is the central data cloud that appears as noises including somewhat signals as revealed in the later chapters. The spines radiating from the central data cloud display obviously signals indicating the large seismicity rates. If the spines of data cloud are elongated toward the directions among the coordinate unit vectors, it shows that there are clear correlations among seismicity rates of each locality cells. Therefore, the patterns of data cloud in the high dimensional space of all locality cells seismicity are a manifestation of correlated seismicity of the global earth. As shown in Fig. 5.12, the spine patterns of the data cloud of the global earth are good examples for this type of correlated seismicity of NE Japan, Tonga, and Sumatra in the sense of naked eyes. In order to obtain the quantitative patterns, the dimension reduction methods in the sense of data-driven sciences should be necessary as is shown in the later chapters.

In the case of the data matrix of Japanese islands region, the spines are clear in the 3D diagram of seismic rates of crust and mantle in cells of Kanto, Ryukyu, and NE Japan as shown in Fig. 5.13. The figure indicates that there are two spines of NE. Japan and Ryukyu and NE. Japan and Kanto. It should be noted that there is a special spine elongated toward the NE. Japan but it means the co-seismic and post-seismic activity of 3.11 at 2011 giant plate boundary earthquake. In this diagram, the spines of the correlated seismicity and individual large co- and post-earthquake seismicity appear the similar adjacent peaks in the data cloud.

The data cloud has much information of time increments during the time scales of the investigation. As introduced in the previous sections, the data cloud is the point

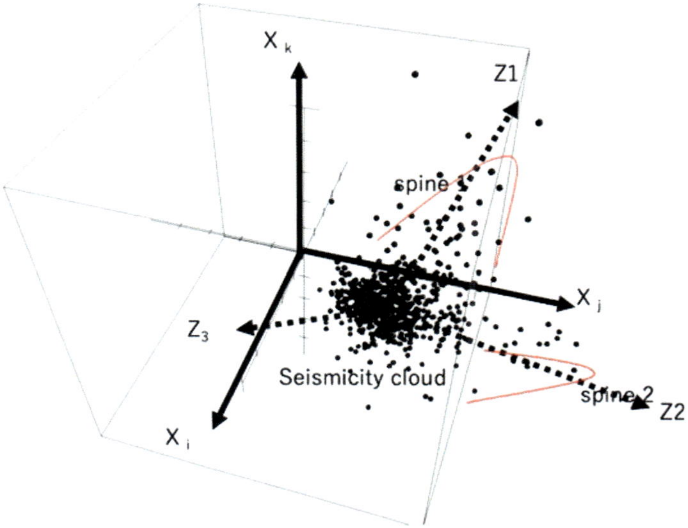

Fig. 5.12 Three dimensions representation of the logarithmic seismicity cloud showing dense core and spines of high seismic activity. X means logarithmic seismicity of locality cell and Z does the correlated seismicity rates taken as principal basis vector toward the spine

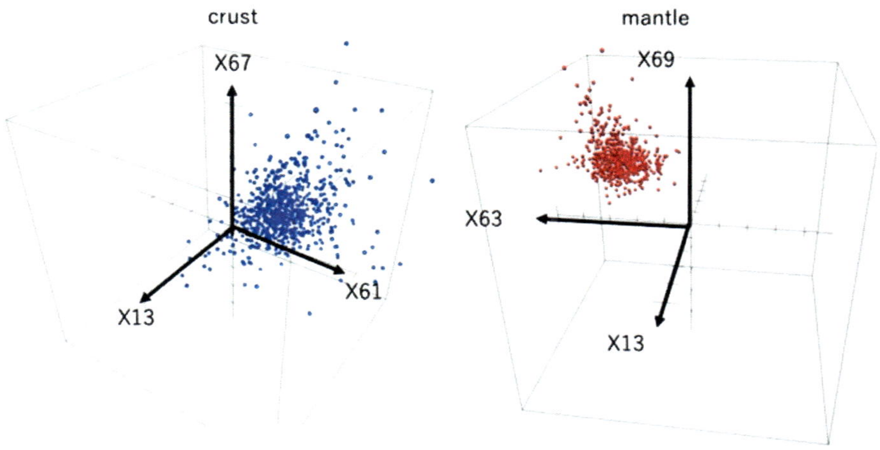

Fig. 5.13 Three dimensions representation of the data clouds for the crust and mantle of the Japanese islands region. The numbers of the logarithmic seismicity rates show the number of locality cell

set of time series of seismicity rates, and every points are connected essentially with each other. So it seems that the spines of the data cloud are the potential indicators of the seismicity dynamics if they are evolved from one to another with increasing the time steps, and thus the tie lines between neighboring time steps of the spines are

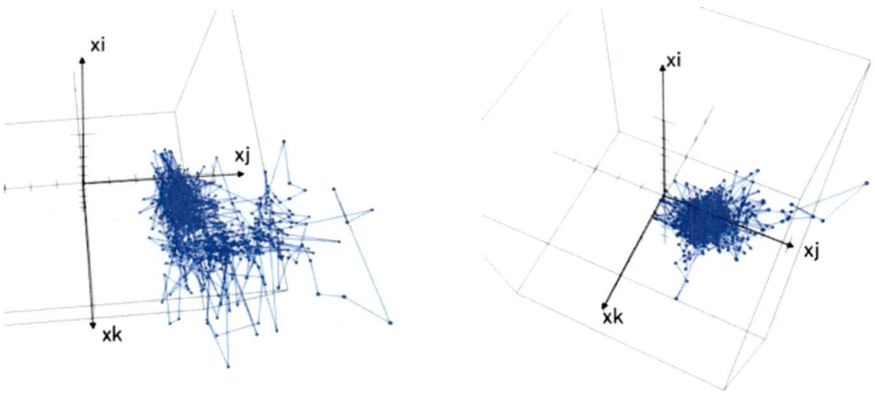

Fig. 5.14 Data clouds with connection of data points showing the two patterns of spines: the left is the progressive spine extension but the right is the isolated spines

shown in Fig. 5.14, suggesting a possibility of systematic change from one spine to another spine in this figure.

The 3D sections of data cloud are shown in the cases of the upper mantle beneath the Japanese islands region in Fig. 5.13. The typical features of the data cloud display several spines showing the correlation patterns of seismicity of Tohoku, Ryukyu, and central Honshu cells and the time track from one to another spine. This time track indicates the transition from one correlated seismic activity to another correlated one, suggesting the spatiotemporal change of the regional seismicity in the Japanese islands region. In the above case, the progressive change from the correlation of Tohoku and Ryukyu to that of Tohoku and central Honshu. The more detailed analyses of this transition of the correlated seismicity should be discussed in the result of the big data analyses by PCA and sparse PCA in the later sections.

5.2 Data-Driven Science of Global Seismicity Dynamics

The data matrix of the seismicity rates in the global subduction zones and the spreading centers are listed in electronic supplementary materials (ESM) for some time windows. In order to study the behavior of the correlated seismicity rates in the whole cells of the subduction boundary in the world, the high dimension data matrix composed of 174 cells (subduction zones) and 168 cells (mid-oceanic ridges) are to be processed by means of two methods: One is the deterministic PCA and the other is the sparse PCA with machine learning by L^2 norm of transformed data and L^1 norm of necessary basis vectors. The primary data of seismicity rates in every cell are transformed to logarithmic seismicity rates as defined in the previous sections, and data matrix composed of about 920 rows and 174 columns (165 columns in the ridges) in the case of 10 days window. In the case of 5 days window the number of rows is about 1840, but it is 9200 in the 1 day window from 1990 to 2018 for the data.

The deterministic PCA analysis of the data matrixes is the method that the data cloud in the primary basis vector space in terms of many cells is projected in the new basis vectors space composed with eigenvectors of the covariant matrix of the primary data matrix as illustrated in Fig. 5.7. This means the transformation of data components, X by the unique rotation orthogonal basis matrix W into Z as

$$Z = W(X - X_m) + N(0, \sigma^2) \tag{5.4}$$

where X_m is the mean value of X, and N is the Gaussian noise term having mean value = 0 and variance = σ^2.

As defined in the previous section, the data matrix x_j can be defined by logarithmic transformation of the volume-normalized raw seismicity rate n_j as following;

$$x_j(t) = \log(1 + n_j(t)/a_j) \tag{5.5}$$

where a_j is the volume normalization factor defined by

$$a_j = \cos(\pi L_j/180) \tag{5.6}$$

in which L_j is the latitude of studied jth locality cells.

In order to determine uniquely W which makes the maximum covariance orientation of the data matrix under the rotation of orthogonal matrix, the following evaluation function should be used (Bishop 2006);

$$J = u\,Su^{\mathrm{T}} + \lambda\left(1 - uu^{\mathrm{T}}\right) \tag{5.7}$$

In which S is the covariance matrix (symmetric matrix) of the data showing $\{s_{ij}\}$, and u is the basis vector of the rotated orthogonal coordinate, λ is the Lagrange parameter, and u^{T} is the transposed vector of u as follows;

$$s_{ij} = (1/N)\sum_{k=1}^{N} x'_{ik}x'_{kj} \tag{5.8}$$

where N is the number of total samples.

Here, it holds

$$x'_{ik} = x_{ik} - x_{im} \tag{5.9}$$

and,

$$x_{im} = (1/N)\sum_{k=1}^{N} x_{ik} \tag{5.10}$$

where N is 754 in the case of 10 days window.

This type of evaluation function under $uu^T = 1$ can satisfy the maximum variance by taking

$$\mathrm{d}J/\mathrm{d}u = 0 \tag{5.11}$$

and then we have

$$S\,u = \lambda\,u \tag{5.12}$$

In this scheme, the highly correlated seismicity changes in the global plate boundaries regions are able to obtain straightforwardly by the projection of the data on the eigenvector direction having the largest eigenvalue, and secondly correlated seismicity by that on the second eigenvector direction, and so on.

In the projected data on the new coordinate basis vector defined by eigenvectors of symmetric covariance matrix from data matrix, the average data point in the original high dimensional data space is plotted at the origin point of the new coordinate system and thus some of transformed data points on the new coordinate may take the minus values which depends on the direction of the new basis vectors as illustrated in Fig. 5.12. In this case, the deviation from the origin point of the new coordinate displays the change from the average seismicity rate as the same as the positive deviation of the transformed data points.

The dimension reduction of the PCA data matrix is performed by the evaluation of the summation of eigenvalues of covariance matrix of data. The eigenvalues decrease rapidly with decreasing the order of the eigenvalues, and the sum of them also decreases rapidly with summation number of them as shown in Fig. 5.15. Thus, the effective number of the eigenvectors for enough presentation of the original data cloud is not the full number of the rank of the covariance matrix, but it commonly becomes enough less than that rank and it is called as the rank of the principal subspaces. In this book, the author intends to show that the seismicity data matrix of the global and regional scales which will be discussed in the later chapters are possibly transformed into principal subspaces having 6–8 dimensions. Higher components of transformed data appear nearly Gaussian random noises with mean value of zero. Therefore, it is obvious that the transformed data variations should have the relatively large random fluctuations in their time series.

Let us look at the synthetic data matrix involving random time series and correlated variations of the signals as shown in Fig. 5.16. The cells of the synthetic seismicity rates have 20×20 matrix and time steps of the simulated data of the cells are counted to be 1000. Then, the synthetic data matrix shows the 400×1000 components. The configuration of the correlated seismicity cells is aligned vertical in the cell column and uncorrelated seismicity cells are aligned in the horizontal cells, but the random noises cells are occupied by other cells.

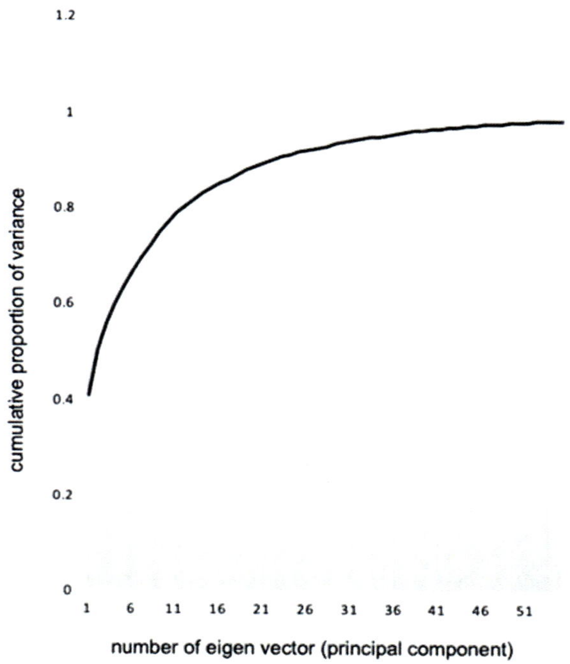

Fig. 5.15 Diagram showing increasing cumulative proportion of variance determined by eigen-values of data matrix

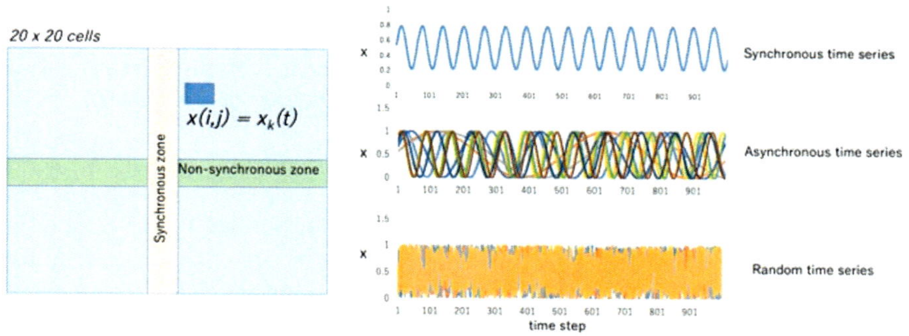

Fig. 5.16 Illustration showing the configuration of the simple synthetic data matrix composed with synchronous, non-synchronous, and random artificial seismicity variation x(i,j). Right, synthetic three types of seismicity time series; one is the synchronous, second the non-synchronous, and third the random patterns

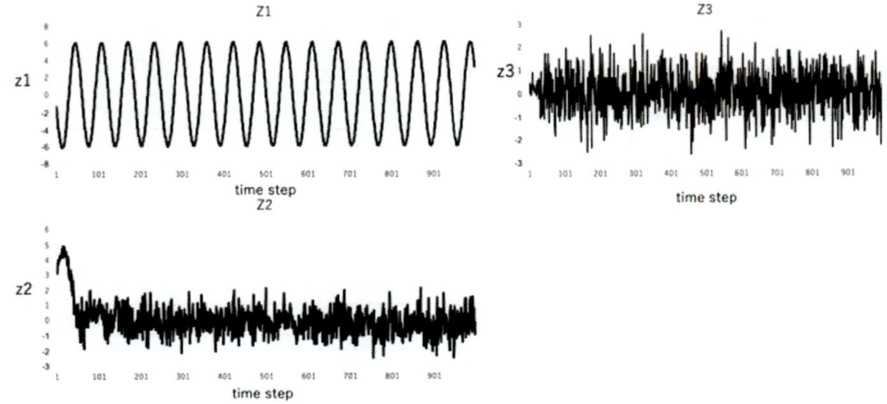

Fig. 5.17 Time series of the main correlated synthetic seismicity rates, showing that the z_1 indicates the synchronous seismicity rates, the z_2 and z_3 do the non-synchronous and random ones

The 400×1000 data matrix of the synthetic seismicity of whole cells are processed to obtain the characteristic features of the time series of the strongly correlated seismicity variations and uncorrelated ones with large random noise by means of deterministic PCA analysis, and the results are shown in Fig. 5.17. It is obvious that the strongly correlated periodic patterns behave the same as the originally given variation pattern and the uncorrelated periodic patterns do the several different variations with large random noises, suggesting the modulations of some periodic variations. The localities (cells) showing the strongly correlated seismicity and partly correlated one changes are exactly reconstructed as the given original configurations as shown in Fig. 5.18. As a result, it is safely concluded that the deterministic PCA analysis can be available for the reconstruction of the strongly correlated time series in the regionally mapped noise and signal sources.

The simple explanation of the PCA processing of high dimensional time series seems such that the original data cloud in the primary coordinates is well explained by the rotated orthogonal space which is uniquely transformed by the orthogonal rotation matrix defined by eigenvectors of covariance matrix of data matrix. In this processing, the first component projected on the first eigenvector with largest eigenvalue display the important main pattern of the time series of spatially correlated variations although the second and third components show the modulations and synthesized correlated time series of the original data cloud. The randomly fluctuated patterns are transformed into random noise patterns too by this method. Furthermore, the higher components than fifth of PCA are needed for inference of the dynamics of the system because of generation of many modulated modes in the case of slightly correlated periodic variations in the regions.

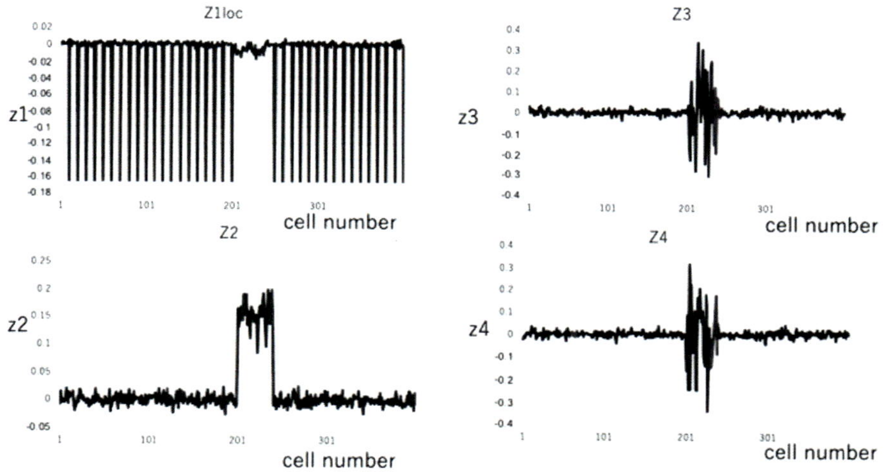

Fig. 5.18 Linear map of correlated locality cells with correlated seismicity rates z_1, z_2, z_3, and z_4 of the synthetic seismicity time series

5.3 The Characteristic Features of Strongly Correlated Global Seismicity

The seismic activity of the solid earth is characterized by the global patterns consisting of strong activity along the subduction zones and relatively weak one along the mid-oceanic ridges. The seismicity rates ratio between their activities reaches about 25 as shown in Fig. 5.5. Thus, the data processing should be made separately on each global data sets of subduction zones and spreading center zones.

First, the strongly correlated seismicity rate patterns are shown for the data of global subduction zones in Fig. 5.19. The trend of the z_1 component time series appears a pattern of step-like increase from 1990 to 2010 and then periodical change from 2010 to 2018. The residual variation of z_1 trend as shown in Fig. 5.20 bottom appears a weak annual to biannual variations during 2010 and 2018 but it disappears before 2010. It seems very important that the z_1 trend shows a gradual increase signals before 2004 Sumatra giant earthquake but an abrupt decrease of z_1 component is seen at 3.11 of 2011 Tohoku-Oki giant earthquake. Furthermore, the residual z_1 of short-term signals appear the peak of annual to biannual variations after 2011 Tohoku-Oki giant earthquake (Fig. 5.20).

The second component of PCA transformation of the global seismicity data matrix of subduction zones is characterized by the gradual increase from 1990 to the data of 2004 and then turns to decrease toward 2011 and again increases to 2017. The variation of the last phase from 2017 to 2018 shows the weak decrease. The secular trend of the z_2 variation and short-term variation of z_2 are displayed in Fig. 5.21, suggesting that the sharp minimum point near 2011 Tohoku-Oki earthquake and gradual decreasing pattern before 2011 and increasing pattern after 2011. The latter

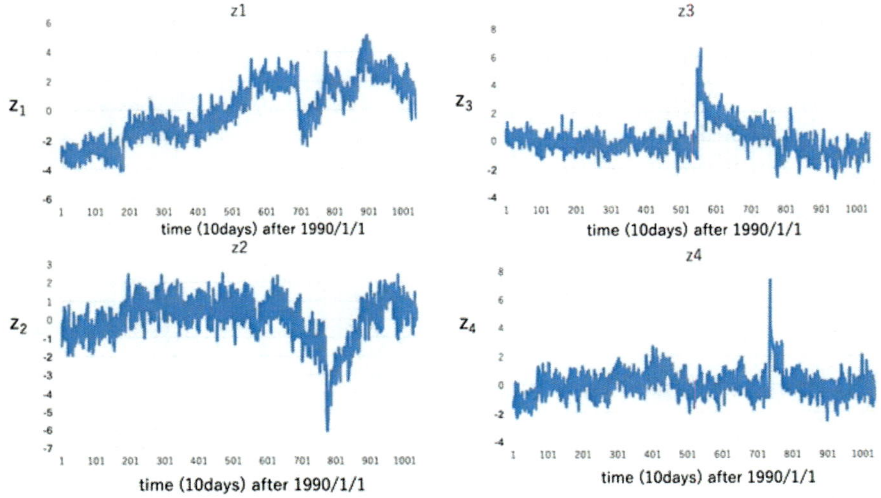

Fig. 5.19 Time series of the main correlated seismicity rates of the global subduction zones showing the secular continuous change and discontinuous rapid change. Horizontal axis is the time in 10 days unit and vertical axis is the correlated logarithmic seismicity rates minus average one

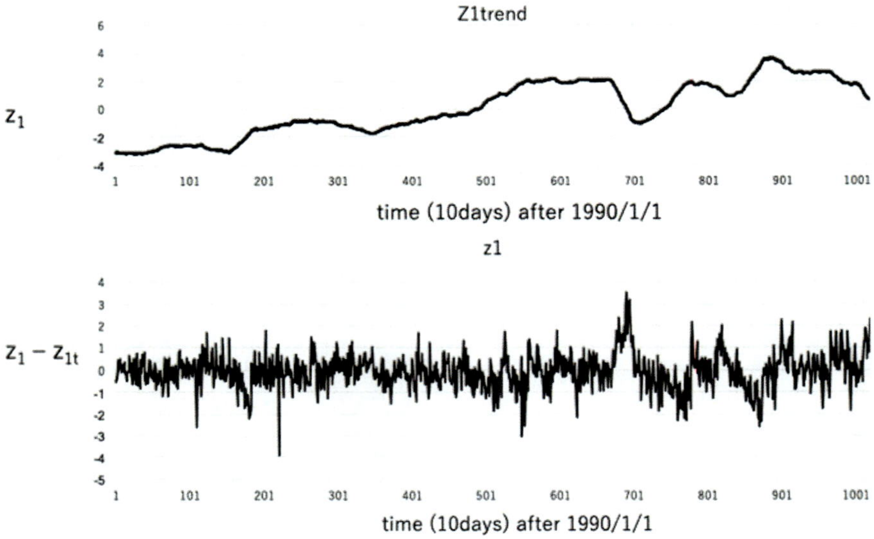

Fig. 5.20 Time series of the secular trend of the global correlated seismicity rate z_1 by migrating average and short-term variation (bottom) of the z_1 by subtraction of the secular trend z_{1t}

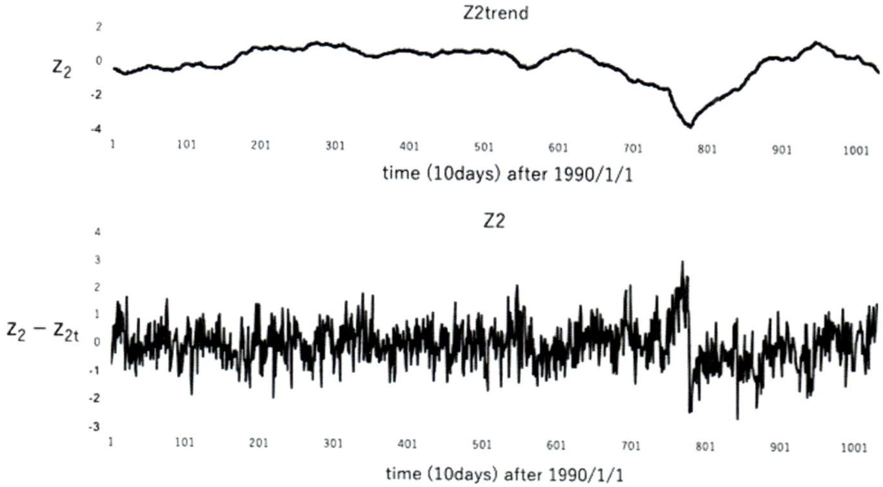

Fig. 5.21 Time series of the secular trend of the global correlated seismicity rate z_2 by migrating average and short-term variation (bottom) of the z_2 by subtraction of the secular trend z_{2t}

pattern is probably responsible for the recovery stage, but the former shows probably foregoing signals of the seismicity change. The gradual increase in z_1 toward 2004 also suggests the foregoing signals before 2004 Sumatra earthquake. It may appear that the short-term fluctuation pattern of z_2 has weak peaks slightly before 2004 and 2011. The annual and biannual variations are not clear in the short-term fluctuation of z_2 as shown in Fig. 5.21.

The third term of z_3 trend shows the simple patterns consisting of slow decreasing, and rapid jump at 2004 Sumatra earthquake, and then gradual decrease after post-seismicity after Sumatra earthquake, although there is a small minimum at 2011 as seen in Fig. 5.22. The short-term fluctuation displays nearly random noise and subtraction effects near the 2004 minimum of the secular trend mentioned above.

The fourth term of z_4 displays somewhat little periodicity and the time scales of the periodicity are about 3–5 years as seen in Fig. 5.23. The sharp peak near 2010 in Fig. 5.19 is probably the post-seismic activity of Java trench.

The fifth and higher terms of z time series appear almost always Gaussian noises have mean value of zero but various variances. The histograms of noisy fluctuation patterns of z_5 and z_6 are seen in Fig. 5.24, suggesting a good approximation of Gaussian distribution of noise. These noisy patterns resemble those of short-term fluctuations except for annual and biannual variations appeared in z_1 and z_2 described previously.

The components of eigenvectors for z_1 to z_4 are shown by the intensity spectrum as indicated in Fig. 5.25. The components of eigenvectors are the coefficients of transformed base vectors (u_i) from original ones (a_i) and expressed as

$$u_i = w_{i1}\,a_1 + w_{i2}\,a_2 + w_{i3}\,a_3 + \cdots + w_{in}\,a_n \quad for\ z_i \tag{5.13}$$

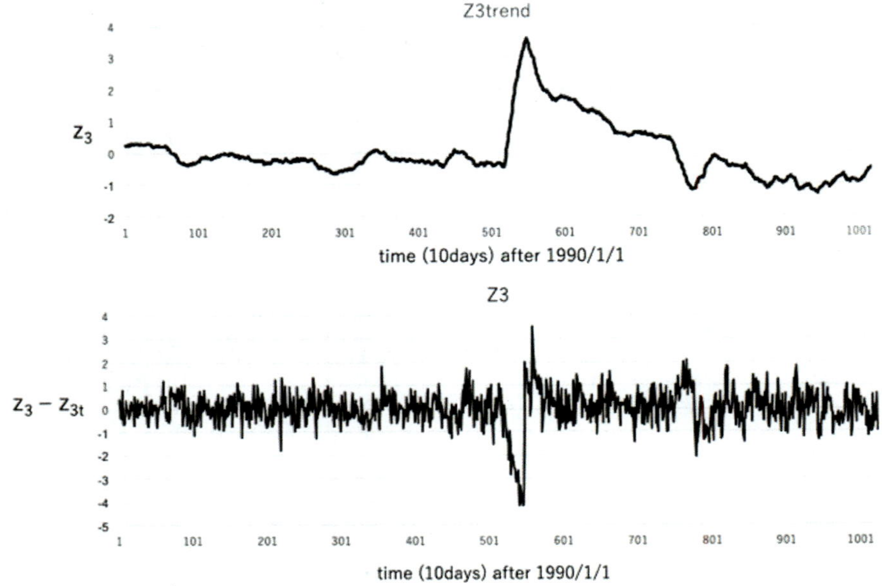

Fig. 5.22 Time series of the secular trend of the global correlated seismicity rate z_3 by migrating average and short-term variation (bottom) of the z_3 by subtraction of the secular trend z_{3t}

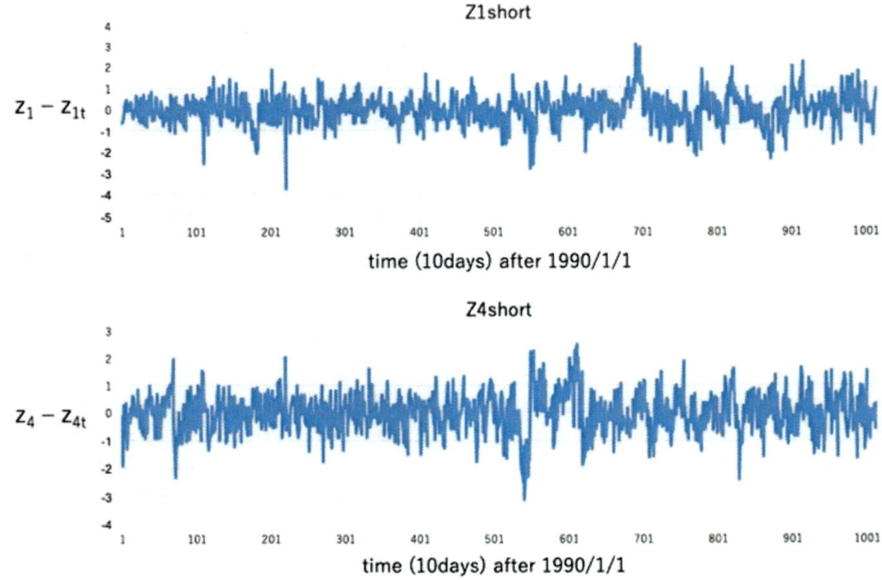

Fig. 5.23 Comparison of the short-term variation between z_1 and z_4 in the global subduction zone showing the rather periodical change in z_1 but random one in z_4

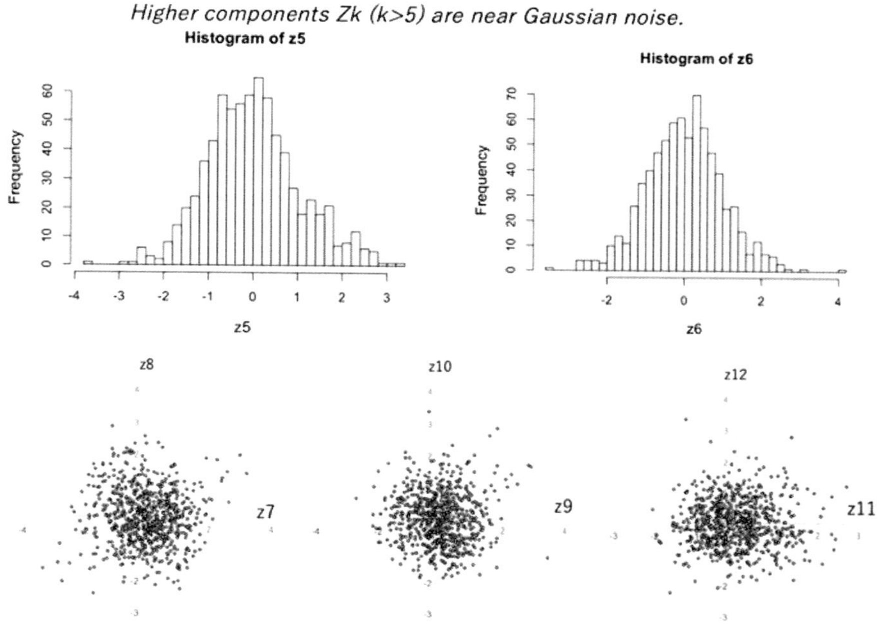

Fig. 5.24 Frequency diagrams of z_5 and z_6 (upper) and higher z components (bottom) correlated seismicity rates in the global subduction zones, showing the Gaussian type

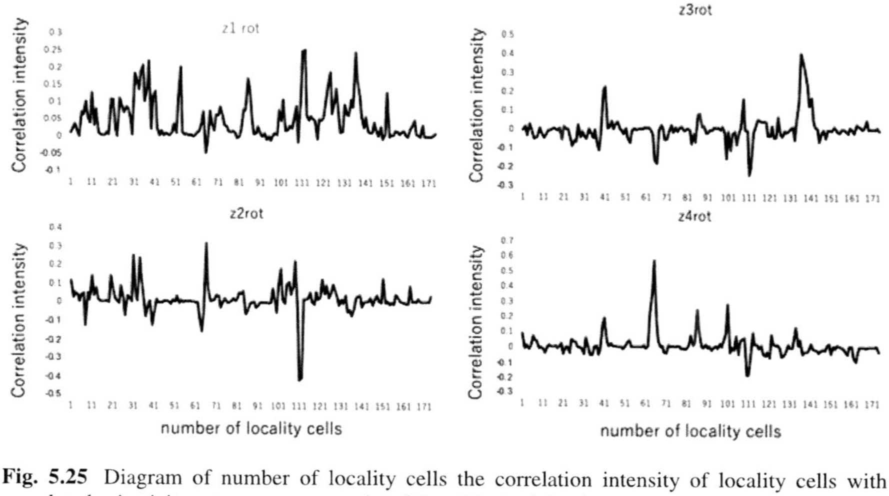

Fig. 5.25 Diagram of number of locality cells the correlation intensity of locality cells with correlated seismicity rates z_1, z_2, z_3, and z of the global subduction zones

in this equation original base vectors a_i are corresponded to the locality cells defined in the previous chapter. Thus, the coefficient w_{ij} means the weight factor for the contribution rate of the local seismicity in the ith cell. Therefore, in the book, the author would like to call the diagram of w_{ij} and cell number j as the intensity spectrum of the z_i and it displays the characteristic strongly correlated locality pattern of the global and regional seismicity.

The intensity spectrum (Fig. 5.25) of the z_i appears multiple peaks at various locality cells listed in Appendix. Some of peaks of z_1 are commonly appeared in z_2, z_3, and z_4 but others are not seen in them. This is the reason why there are different segments in the single long time series of every z_i in terms of the characteristic pattern of the secular trend and why the correlations between each segments of these time series become significantly large. Large weight factors of these common segments seem to be come from the post-seismic activity of the giant earthquakes, for example, Sumatra 2004 and Tohoku-Oki 2011 giant earthquakes as being identified by the common peaks at their locality cell numbers in Fig. 5.25.

The intensity spectrum of z_1 is characterized by the strong seismicity correlation among Tohoku-Oki, Tonga-Kermadec, Chile, Aleutian, Mindanao, and Sumatra, but on the other hand that of z_2 by that among Tonga- Kermadec, Chile, and Tohoku, and that of z_3 by Solomon, Tonga-Kermadec, Tohoku, and Sumatra. Therefore, the z_1 manifests almost always strong correlation of global subduction regions of Pacific plate, Indian Ocean plate, Antarctica plate, Africa plate, and Eurasia plate.

The global patterns of the strongly correlated seismicity regions are possibly seen in the global maps using the Google Earth as shown in Fig. 5.26. The patterns of the z_1 components of seismicity appear the correlated activity of western Pacific and eastern Indian Ocean regions of the subduction zones. On the other hand, the patterns of the z_2 components display the interesting mode of correlated seismicity that the seismicity rate of the central Chile is strongly synchronized with that of the northeastern Japan and northeast of the Hokkaido regions. As noted in the later sections, the secular trends of the z_1 and z_2 look like the anti-phase synchronization mode as shown in Fig. 5.26. It means that the global mode coupling between southeastern Pacific and northwestern Pacific subduction zones occurs at the northeastern Japan region and it seems to be a keystone of the anti-phase seismic activity of the southeastern and northwestern Pacific subduction zones mentioned above.

These different intensity spectra of z_i should influence against their time series patterns. Let us consider the schematic relations between the intensity spectra and the time series of z components as follows:

There is an incompatibility between z_1 and z_2 or z_2 and z_3 intensity spectra as shown in Fig. 5.27, and there is an anti-phase synchronization between their time series. At this case, it seems that the seismicity connecting two locality cells appeared in z_1 intensity spectrum changes another configuration connecting different locality cells repeatedly as illustrated in Fig. 5.27. On the other hand, there may be a different situation that the seismicity connecting three locality cells (mode I) and seismicity connecting ten locality cells involving above (mode I) locality cells (mode II) have the different time series with common segments as mentioned previously. The latter case is likely to the model shown in Fig. 5.28, suggesting that the global seismicity

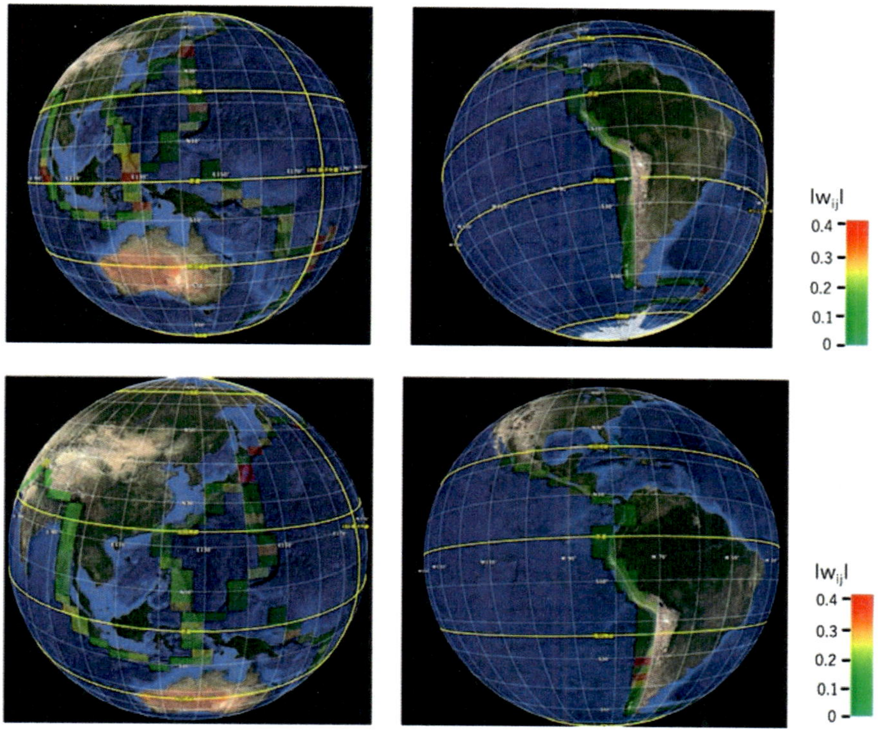

Fig. 5.26 Map of the correlation intensity of the locality cells with the correlation seismicity rates in the global subduction zone on Google Earth. The intensity of the correlation degree is shown by pseudocolor. The upper is z_1 and bottom is z_2

starts with several local activities and then the global seismicity changes to correlated activity of widely distributed local cells.

The case of the actually global seismicity correlations of z_1 and z_2 is likely to the case of the latter: The very widely correlated seismic regions continued a gradual increase in activity from 1990 to 2003 and the small number of correlated seismic regions already attains a maximum level before 2002. The second term mode of seismic connections decreases after 2002 and then the first term mode keeps the high-level seismic activity still now. During the latter phase, the giant Tohoku-Oki earthquake occurred at the lowest level in the z_2 time series (Fig. 5.19).

These changes of the seismicity activity of correlation modes mentioned above should be interpreted more clearly by the attractor diagram of Figs. 5.29 and 5.30. The attractor diagram is shown by the z_1 versus z_2 or z_2 versus z_3 diagrams with increasing the time advance. All data of z_1, z_2, and z_3 are plotted in the attractor diagrams but in order to understand the trajectory of the attractors should be investigated by the successive change of the data plotting with several time intervals on the diagrams. In this study, the author intends to divide four time intervals of 1990–2009 and four

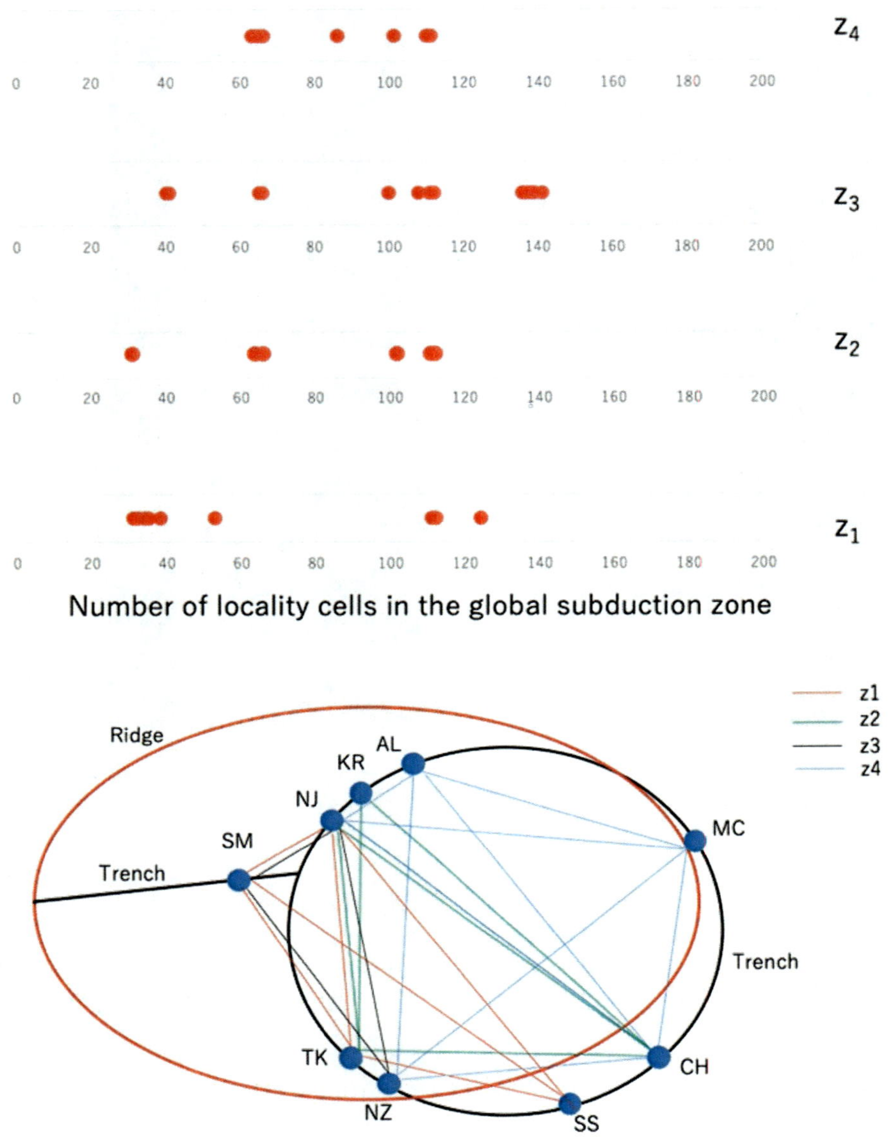

Fig. 5.27 Illustrative diagram of the representative correlated cells of the correlated seismicity rates z_1 to z_4 in the global subduction zone. Red dots show the representative correlated locality cells indicating the correlation network. The bottom figure displays the modes of network structure in z_1 to z_4. Bottom figure: AL; Aleutian, KR; Kuril, NJ; Northeast Japan, SM; Sumatra, TK; Tonga-Kermadic, NZ; New Zealand, SS; South Sandwich, CH; Chile, MC; Mexico

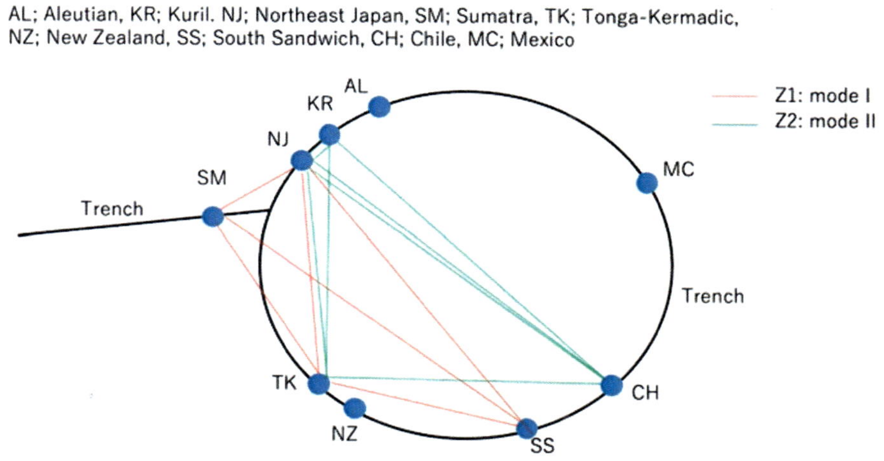

Fig. 5.28 Representative correlated seismicity network of mode I and II in the global subduction zone. AL; Aleutian, KR; Kuril, NJ; Northeast Japan, SM; Sumatra, TK; Tonga-Kermadic, NZ; New Zealand, SS; South Sandwich, CH; Chile, MC; Mexico

Fig. 5.29 Three dimensions representation of the correlated seismicity rates in the global subduction zones, showing tie lines between plots of neighboring data. The dense clusters A, B, and C, and scattered spines at 2004 and 2011 can be identified

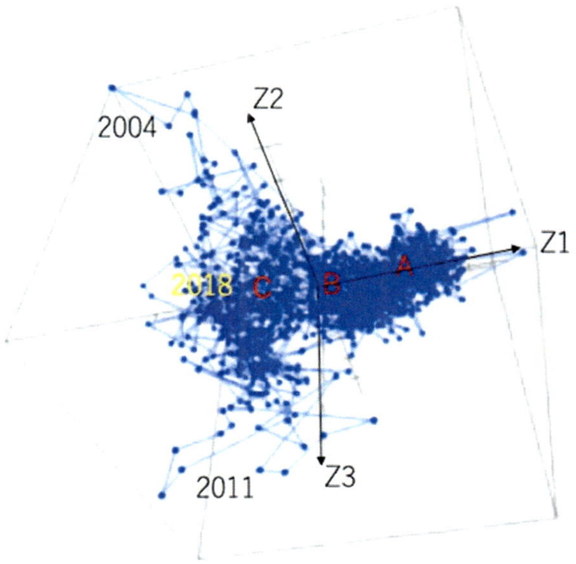

intervals of 2009–2018 as shown in Fig. 5.31. It is obvious that data of every interval make different sharp clusters with random Gaussian noises and they systematically change from low z_1 to high z_1 regions, suggesting the global seismicity change from low seismicity to high seismicity activity.

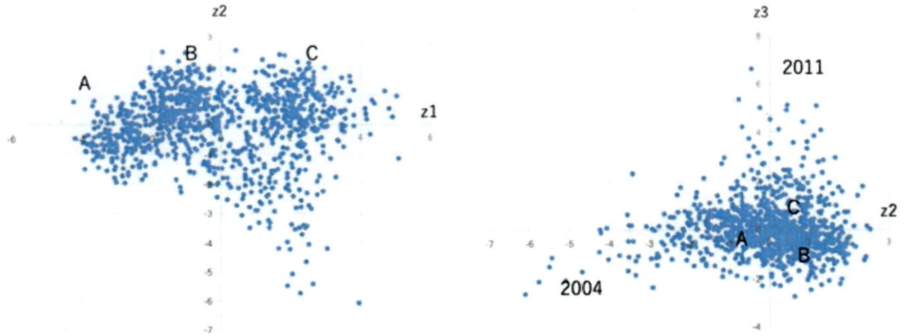

Fig. 5.30 Diagrams of the z_1 and z_2, and z_2 and z_3 of the global correlated seismicity rates of the 10 days window in the subduction zones from January 1, 1990—June 12, 2018. A, B, C and 2004 and 2011 are the same as Fig. 5.29

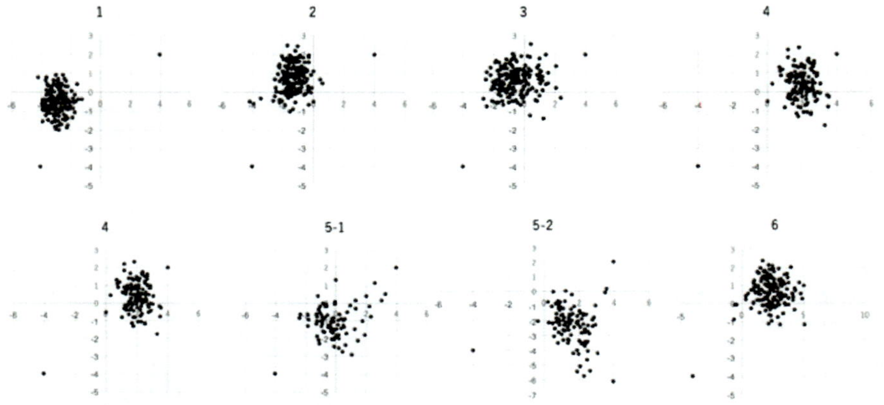

Fig. 5.31 The z_1 (horizontal axis)–z_2 (vertical axis) diagrams of the progressive change of the correlated seismicity clusters with time in the global subduction zones from 1990 to 2018. Upper; 1990 to 2009, bottom; 2009 to 2018

It should be noted that the variance of the clusters changes somewhat large before the giant plate boundary earthquakes: In the case of preseismic activity of the Sumatra earthquake, the variance along the z_1 components changes from 0.3 to 0.6 and also of preseismic one of the Tohoku-Oki earthquake it does from 0.5 to 1.0 along the z_2 components. In addition, the relationship of the variance between z_1 and z_2 seems to be classified by two types: One is dominant of variance change in z_1 and another in z_2 as shown in Fig. 5.32. It suggests that the fluctuations in the correlated seismicity tend to become large before the large earthquake in the global subduction regions.

On the other hand, how is the relationship between the mean value of the correlated seismicity and their variance intensity? Figure 5.33 shows the fact that the central value of the mean z_1 increases with increasing the intensity of their variances, in

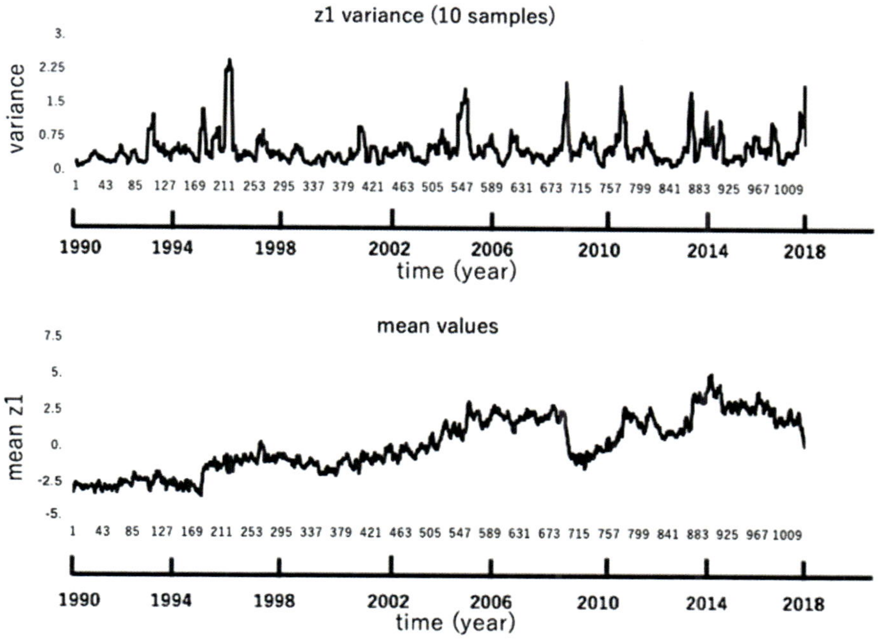

Fig. 5.32 Time series of the variance (upper) and the mean value (lower) change of the correlated seismicity z_1 in the global subduction zones from 1990 to 2018

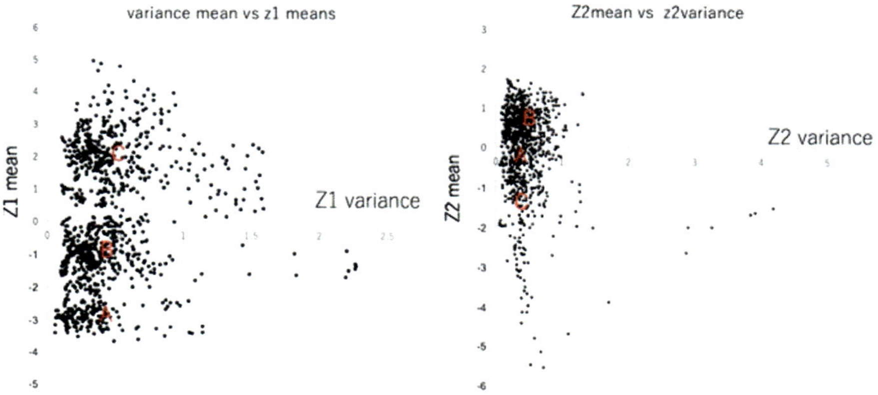

Fig. 5.33 Diagrams of the mean correlated seismicity and the variance of z_1 and z_2 in the global subduction zones. A, B, and C are the same clusters as those in Fig. 5.30

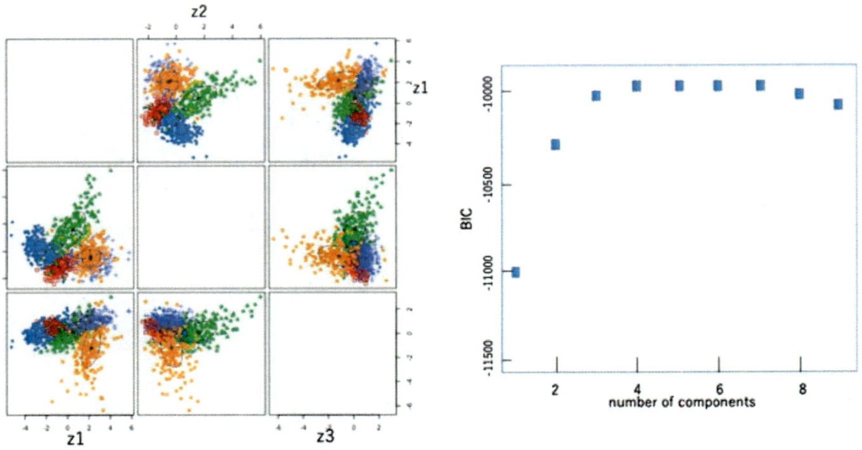

Fig. 5.34 Available clustering of the correlated seismicity rates z_1, z_2, and z_3 and their Bayesian information criterion (BIC) values by "mclust" package of R

spite of the large deviations of the variance intensity appear different patterns of increasing trends with negative slope from the large mean and low variance intensity to the small mean and high variance.

As stated in the previous sections, the seismicity activity before giant earthquakes seemingly involves the large fluctuations suggesting the large variance. Therefore, it looks like that the above-mentioned decreasing trend with increasing variance of the fluctuation is somewhat a signal for the becoming large earthquakes. However, it is already shown that there are two trends toward the giant earthquakes of 2004 and 2011 disasters such as z_1 and z_2 changes. In order to clarify the bifurcation of the z_i trends, it needs that the distinctive clusters of z_1, z_2, and z_3 diagram should be convinced by means of the stochastic methods of the large data sequences of z using the "mclust" packages in R language. Figure 5.34 displays the map of the clusters of data in the z_1–z_2–z_3 diagram and the cross-validation index of BIC for various sizes of clustering models. The maximum BIC model is most available for the clustering of the data in this diagram and the elliptical multiple Gaussian clustering having five clusters is the most robust model.

Then, the five clusters model is acceptable for the whole data set of the global seismicity rates and in this map the clusters are also aligned with advancing the time described previously as is seen in the z_1–z_2 diagrams of which each diagram is divided by the time sequences (Fig. 5.31). As noted in that sections, as the preseismic fluctuations before large earthquakes seem to become large, the fluctuations of z data in the periods of large earthquakes of the 2004 Sumatra and the 2011 Tohoku-Oki are investigated in the above-mentioned cluster map as shown in Figs. 5.35 and 5.36, indicating the distinctive enlarging of the fluctuations in the 2004 and 2011. As is seen in the locality intensity diagram, the z_2 and z_3 components are strongly related to the correlations of NE Japan-Tonga-Kermadec-Chile and Sumatra-Java-New Zealand,

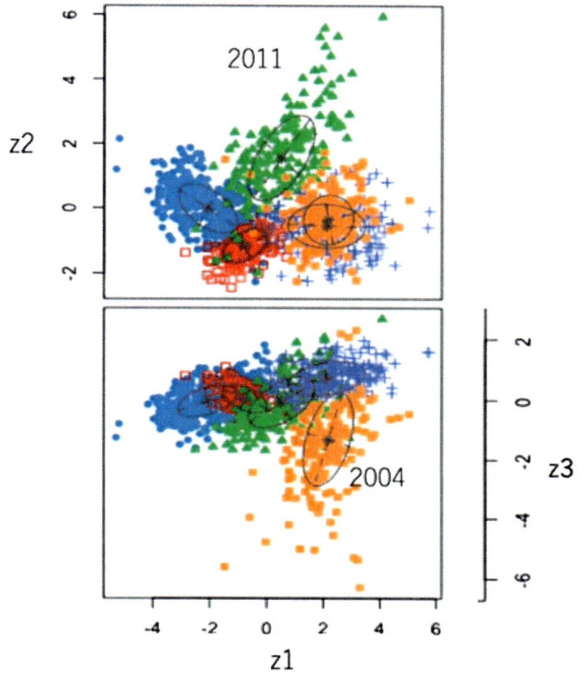

Fig. 5.35 Clusters of the correlated seismicity rates and spines of 2004 (brown) and 2011(green) in the z_1, z_2, and z_3 in the global subduction zones. Blue, red, and purple clusters are A, B, and C clusters in Fig. 5.30

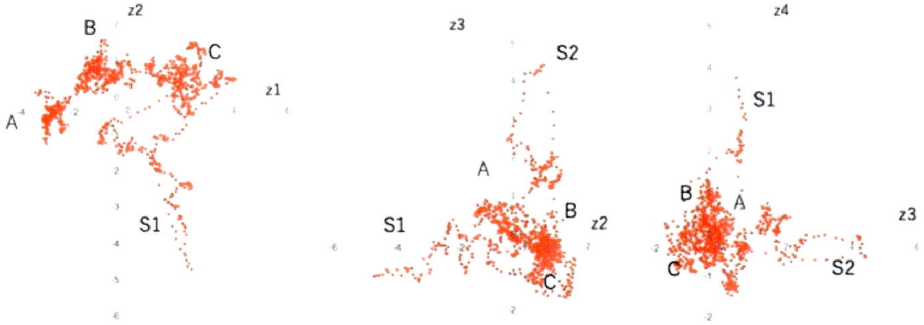

Fig. 5.36 Diagrams of the correlated seismicity rates z_1 to z_3 by averaged values in the global subduction zones. The rather discontinuous clusters in left diagram are seen. Cluster A, B, and C are the same as Fig. 5.30 and S1 and S2 are spines of 2004 and 2011, respectively

respectively, suggesting that the 2004 event is strongly connected with the latter correlated seismic regions and the 2011 is by the former ones.

5.4 Global Seismic Moment Release Rate and Correlated Seismicity Rates

As noted in the previous sections, the global change of the whole seismic moment release rate is represented by the rate of the giant earthquakes in the world because the major contribution of the whole seismic moment is those of the giant earthquakes over magnitude 8. The whole moment release rate normalized by the moment of the earthquake with magnitude 7.5 should be represented by the following;

$$M_t = \Sigma_i \, 10^{1.5\,(m_i - 7.5)} \tag{5.14}$$

where M_t is the total moment release rate normalized by M7.5 per one year, and m_i is the magnitude of ith earthquake over magnitude of 7.5 in one year. The relationship between magnitude and seismic moment is followed by Kanamori and Anderson (1975).

Figure 5.37 shows the secular trends of the M7.5 equivalent moment release rate per year and the annual average value of z_1. The z_1 trend clearly displays the increasing pattern from 1990 to 2003 and then it is largely fluctuated in the high level of z_1 after 2003 to 2016 (also to 2018 as noted in the general trends of z_1 in the previous sections). On the other hand, the moment release rate shows the peculiar variations during 2003 to 2005 and 2010 to 2012. These high moment release rates appear during the periods of giant earthquakes of 2004 and 2011. Besides, it seems noticeable that the relations between moment release rate and strongly correlated seismicity intensity are possibly divided into two regions in the z_1 versus M_t diagram (Fig. 5.38): One is the low correlated seismicity activity region with small fluctuation

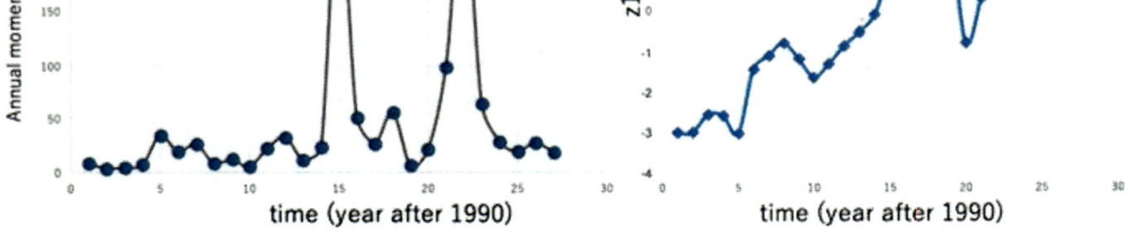

Fig. 5.37 Time series of global moment release rate with M7.5 equivalent value per year (left) and average correlated seismicity rate per year (right) in the global earth

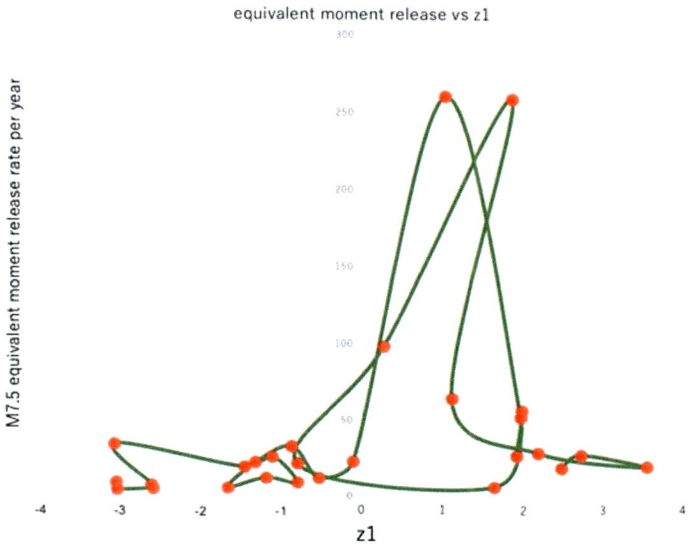

Fig. 5.38 Relation between the global correlated seismicity rate z_1 and M7.5 equivalent moment release rate per year in the global earth

of the moment release rate and the other is the high activity region having very large fluctuation of moment release rate. The former belongs to the period from 1990 to 2003 and the latter to the period from 2004 to 2016 (maybe to 2018). Therefore, it is possible that the global seismic activity appeared in the globally correlated seismicity rate z_1 variation is classified to be the quiet period (1990–2003) and murmuring period (2004–2016) by means of the moment release rate change as suggested by Ekstrom (2007) and Zapliapin and Kremer (2017).

5.5 Correlated Seismicity Rate Variations of Global Ocean Ridges

The Mid-Atlantic Ridge (hereafter called MAR), Mid-Indian Ocean Ridge (IOR), and East Pacific Rise (EPR) are typical ocean spreading centers connected by the transform faults. In the center zones of the ridges, there are high seismicity but the magnitudes of central earthquakes are almost always under six together with seismicity rates less than several percent of those in subduction zones. The seismic source mechanisms in the ridges and transform faults are characterized by normal fault type and strike-slip fault type, respectively.

The thickness of the oceanic plates at the center of the ridges is less than 1 km and it becomes gradually thick to the several tens km apart from the ridge. Underneath the ridge center, there are large partially molten zones having large amounts of basaltic magma and then the temperature of the top of the partially molten zones reaches about 1000 °C. Therefore, the submarine magmatism at the central zones of the ridge is very active erupting the mid-oceanic ridge basalt (MORB) magma.

The seismic activity occurs within the upper several km regions because of the high temperature conditions in the deep regions. As far as the seismicity is responsible for the rock brittleness, the extension of the seismic slip regions is limited in the low temperature regions so that the depth range is under several km. Consequently, the magnitude that is governed by the area of the slip surface seems to be controlled under the value about 6.

The ocean ridges have been characterized by large scale hot spots at the top of the large hot plume rising from the lower mantle regions, for example, the Iceland region in the northern MAR, and Galapagos islands in the eastern EPR. In Iceland, there are abundant rift valleys now spreading the oceanic crust east–westward. In these hot spot regions, there are abundant earthquakes in the crust. As noted in the next sections, the seismic activity of the hot spot crust is very high compared with other regions of the mid-oceanic ridges.

The studied locality cells along the global mid-ocean ridges are from north of Iceland to southwestern South Africa of the MAR, and those of the Indian Ocean Ridge are from south of the Cape Town through the triple joint of central Indian Ocean and to the south of the Australia and New Zealand on one side and on the other side those to the Red Sea rift zone. Besides, the locality cells continue from the south of the New Zealand to the Galapagos Islands and Off Mexico through the west of Easter islands. Basically, the EPR continues to the Baja California through San Andreas Fault to the west coast of Canada, but these segments are composed of large transform faults and very near the continent and trench zones. The total number of the cells along the global ridges attains 168 localities.

As described in the previous chapter, the studied seismic data are the number density of the seismic events in the locality cell per the time intervals (here 30 days). The data sets of 168 locality cells are sorted from the USGS earthquake data center in the range of 1990–2018. The magnitude of present earthquakes is ranged from 4 to 5 for the data processing needs large amounts of seismic events in the single cells for epicenters of earthquakes less than magnitude of 4 show very local distribution near the dense seismic stations regions and display the large discrepancy from the Gutenberg and Richter's law. For the number of seismic events in the ridge centers is very small, the data sets are chosen as those of the time intervals of 30 days in this study. Thus, the map of all locality cells in the global mid-ocean ridges is shown in Fig. 5.1. The raw seismicity rate time series of several locality cells are shown in Fig. 5.39, suggesting the weak activity of the central MAR, western IOR, and EPR, being compared with northern MAR and central IOR. Data cloud of the global ridge seismicity is shown in Fig. 5.40.

First, the whole data matrix of global ridges is processed by the natural logarithmic transformation (one of the Box-Cox transformation) of raw data (number density of seismic events in the single cells) as follows;

$$x_{ij} = \log(1 + n_{ij}) \tag{5.15}$$

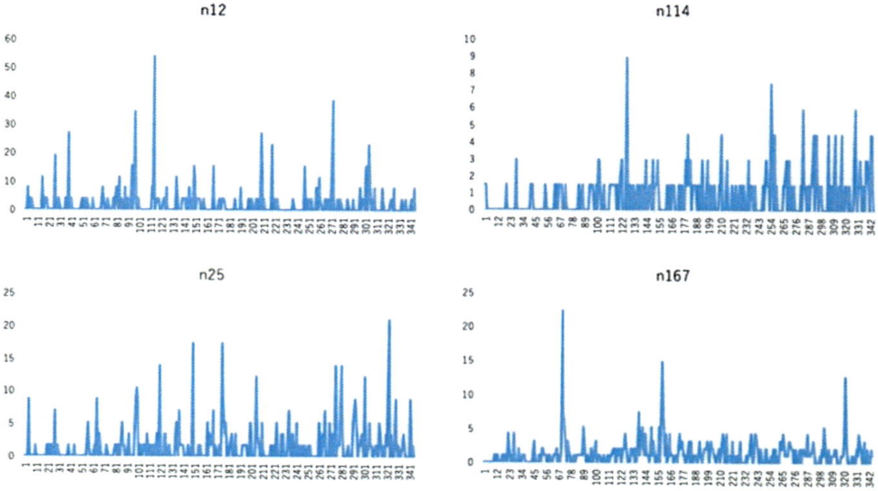

Fig. 5.39 Time series of seismicity rates of representative locality cells (n) in the global oceanic ridge zones. Horizontal axis is the time by 30 days after January 1, 1990, and vertical one the number of seismic events of M4–M5 per 30 days in the locality cell

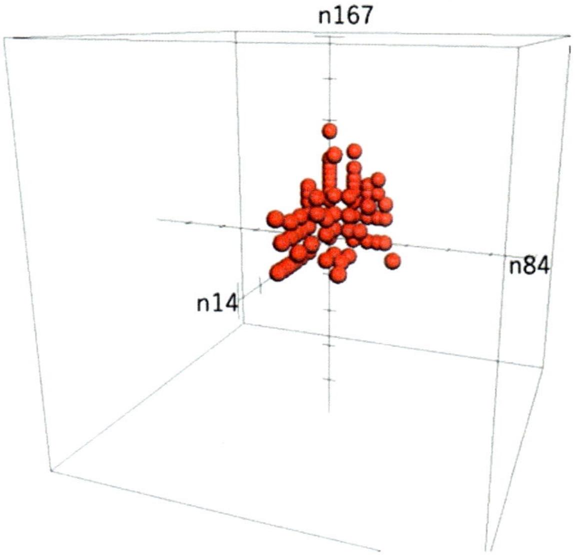

Fig. 5.40 Seismic clouds of the global oceanic ridge from 1990 to 2018 shown by representative locality cells of Mid-Atlantic Ride (×14), Indian Ocean Ridge (×4), and East Pacific Rise (×167)

where x_{ij} and n_{ij} are transformed seismicity rate and raw seismicity rate in the single locality cells, respectively. The total numbers of i and j are 347 of 30 days window from 1990 to 2018 and 168 from MAR and IOR to EPR, respectively.

The large matrix of transformed seismicity rates x_{ij} is investigated by means of deterministic PCA and sparse PCA methods on the R studio installed in the Mac system using R-packages. In this study, the data processing by means of PCA is aimed to obtain the time series of the correlated seismicity regions in the global ridge system, and so that the several components of z_i will be discussed in detail as follows. The time series of z_1 is the most correlated term of the whole data matrix and thus the fluctuation of z_1 will display the characteristic patterns of the basic coherent plate motion at whole spreading centers. Figure 5.41 displays the time series of the z_i for x_{ij}, and the vertical axis means the deviation of z_i from the mean z_i, where $z_i(t)$ can be expressed by the linear combination of $w_{ij}\ x_j(t)$ where w is the weight parameters determined by eigenvectors of symmetric covariance matrix of data $\{x_{ij}\}$.

The general trend of z_1 is possibly identified to be a gradual increase from 1990 to 2018 as shown in the time series and the pattern of the migration average with 5 terms (Fig. 5.42), although the z_2 time series shows the clusters of spike-like peaks

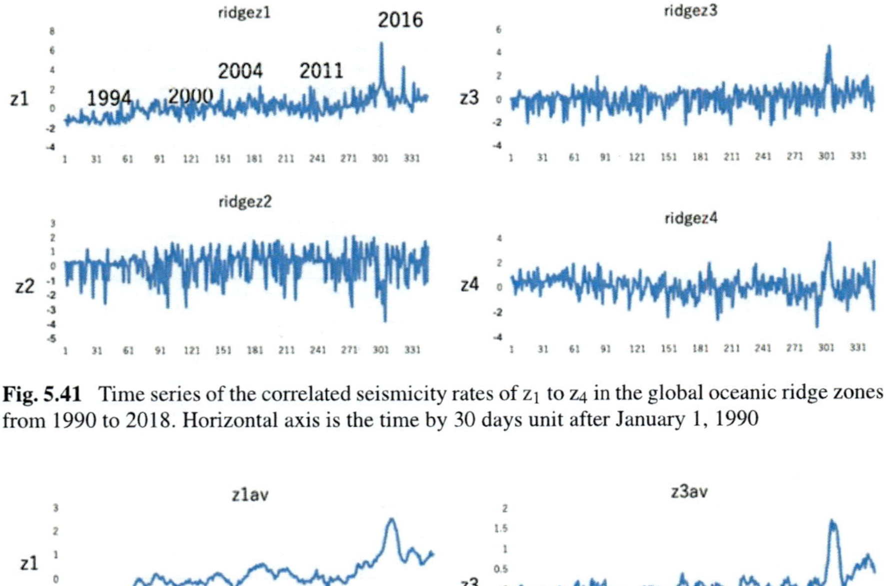

Fig. 5.41 Time series of the correlated seismicity rates of z_1 to z_4 in the global oceanic ridge zones from 1990 to 2018. Horizontal axis is the time by 30 days unit after January 1, 1990

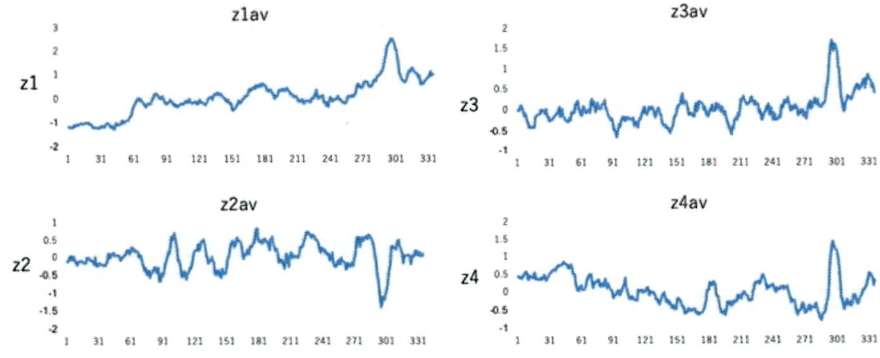

Fig. 5.42 Averaged time series of the correlated seismicity rates z_1 to z_4 in the global oceanic ridge zones after 1990 to 2018. Horizontal axis is the time by 30 days from January 1, 1990

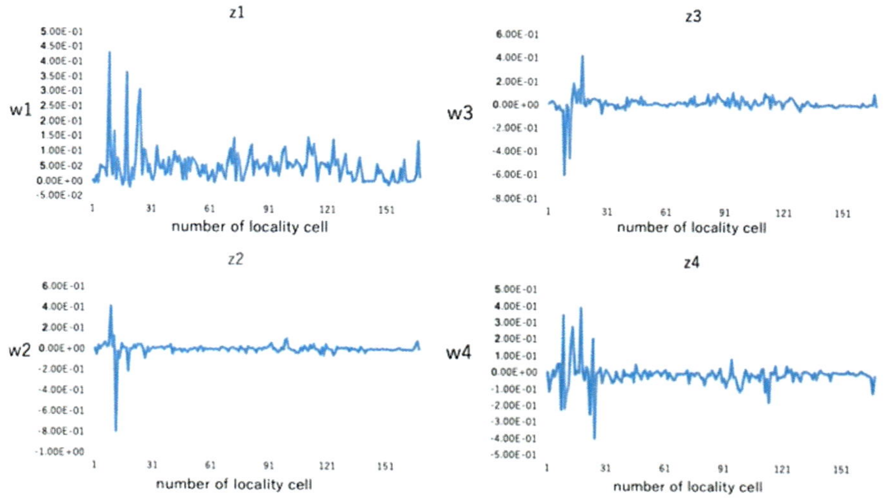

Fig. 5.43 Diagram of correlation intensity of the locality cells of the correlated seismicity rates z_1 to z_4 in the global oceanic ridge zones. Horizontal axis is the number of the locality cells and vertical axis is the correlation intensity w_{ij}

of correlated seismicity rates dominated in Iceland. The spike clusters of correlated seismicity rates, z_3, of the Iceland are also found in the wide range of 1990 to 2018 at the same periods of spike clusters of north of Iceland. On the other hand, the correlated seismicity rate z_1 displays the continuous gentle step-like trend having about 5.6 years periodicity from low to high levels. In the figure, it is noted that the z_1 has the positive major weight coefficients of transformed basis vector as shown in Fig. 5.43 and thus the vertical axis in the z_1 time series diagram shows the upward increase in the correlated seismicity rate. Then, Fig. 5.43 shows that the z_1 of the period of 2005–2018 and especially of 2013–2018 keeps also high level with some degree of fluctuation.

The diagram of z_1 and z_2 shows three clusters: One (A cluster) is the low z_2 regime, the second (B) is the medium z_2 regime, and the third (C) is the high z_3 regime as shown in Fig. 5.44. The general trend of each regime in this diagram displays the flat pattern between z_1 and z_2. The plots on the z_2 versus z_3 diagram as shown in Fig. 5.44 display clearly also three clusters: One is the high in z_2 and z_3, and the second is the high z_2 but low z_3, and the third low z_2 and z_3. The first two clusters show the concentration of plots but third one does a large scattering pattern in the region of low z_2 region. As stated previously, the z_2 and z_3 clusters of spike-like peaks seem to be coeval in the periods of 1996 to 1998, 2000 to 2003, 2005 to 2008, and 2010 to 2013, and then activities of the z_1 become high in the same periods because the z_1 correlated seismicity rates are high in Fig. 5.43 as noted previously. The average plots of z_1 versus z_2, z_2 versus z_3, and z_3 versus z_4 are shown in Fig. 5.45, suggesting that the different clusters appear in the cluster A in Fig. 5.45 and single cluster appears in the C cluster in Fig. 5.44.

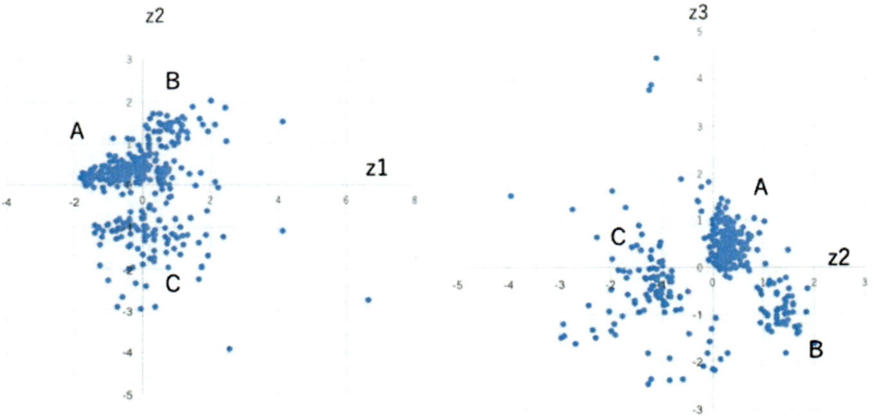

Fig. 5.44 Diagram of the correlated seismicity rates z_1, z_2, and z_3 in the global oceanic ridge zones showing the three clusters of A, B, and C with increasing the time from 1990 to 2018

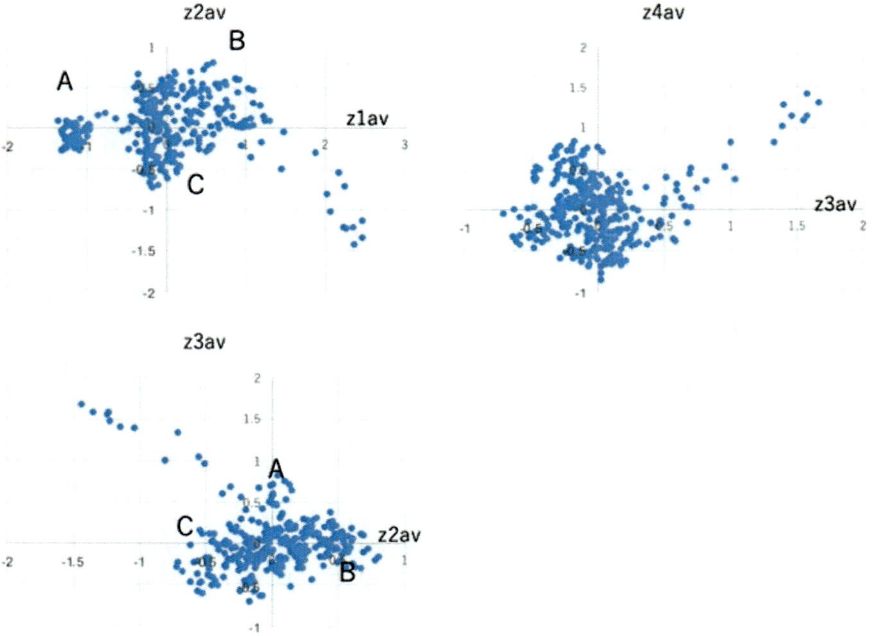

Fig. 5.45 Relations of average values of the correlated seismicity rates z_1, z_2, z_3, and z_4 in the global oceanic ridge. The clusters A, B, and C are the same as those in Fig. 5.45

To investigate the more detail trends of the EPR, IOR, and MAR correlated seismicity rates, the data matrixes of them are independently treated as z transformation by PCA. The results of EPR indicate the weak periodicity of the spike clusters in the z_2 and z_1 time series as well as the global ridge time series of correlated seismicity rates (Fig. 5.46). Considering the positions of locality cells as indicated in Fig. 5.47, the z_1 is mainly composed of the seismicity rates in the locality cells around Off Mexico, and z_2 is of the seismicity rates in the cells of south of Australia. Furthermore, the z_3 is composed mainly of the seismicity rates in south of Australia and

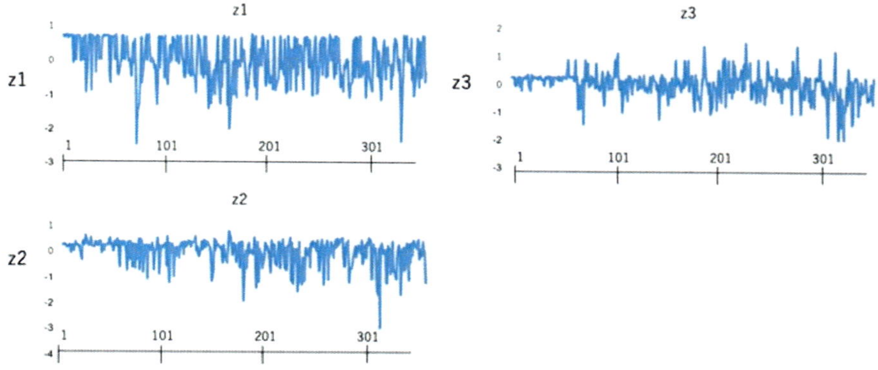

Fig. 5.46 Time series of the correlated seismicity rates z_1 to z_3 in the East Pacific Rise. Horizontal axis is time by 30 days from 1990 to 2018

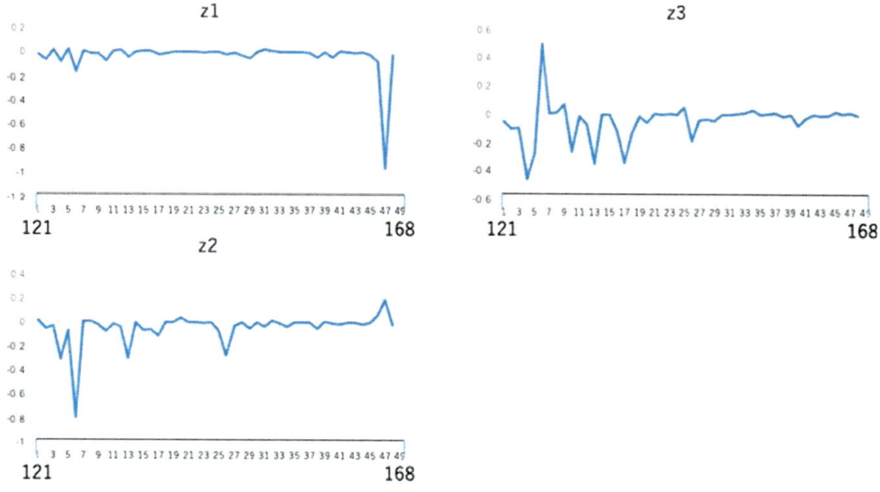

Fig. 5.47 Relations of the correlation intensity of the locality cells in the East Pacific Rise and the correlated seismicity rates z_1–z_3. Horizontal axis is the number of locality cells in the East Pacific Rise

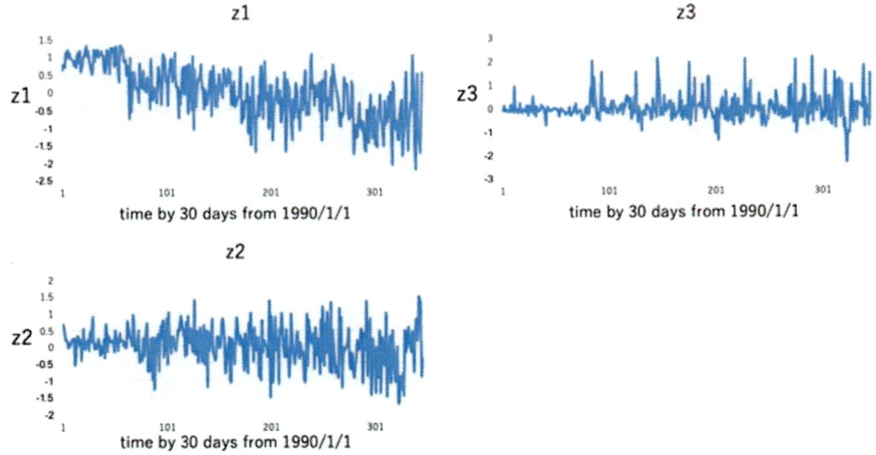

Fig. 5.48 Time series of the correlated seismicity rates z_1–z_3 in the Indian Ocean Ridge from 1990 to 2018

New Zealand. It seems that the time series of the z_2 and z_3 also look like the similar patterns between them though they seem to be weakly periodical of spike clusters.

On the other hand, the z_2 and z_3 time series of IOR display the weak periodic ones of spike clusters similar to those of EPR as shown in Fig. 5.48. The time series of z_1 component appears as a gradual increase having weak periodicity of spike clusters similar to the global z_1 fluctuation. Judging from the diverse contribution of the correlated seismicity rates of the cells along the IOR ridge in the z_1 component (Fig. 5.49), above-mentioned weak periodicity of the spike clusters in the z_1 time series is probably due to the seismicity activity associated with periodical magmatic activity change in the spreading center of the IOR. Such magmatic activity is derived from and/or result in the spreading and plate motion, and thus the global seismicity activity also seems to become coherent with the global ridge activity as discussed in the preceding sections.

Contrasting with the time series of the z components of IOR and EPR, the weight coefficients of locality cells along Mid-Atlantic Ridges are high in the cells of north of Iceland and Iceland as shown in the intensity diagrams (Fig. 5.50). It is probably noted that the z_1 and z_2 time series show the weak periodicity of spike clusters (Fig. 5.51) as seen in the IOR and EPR. It seems that the correlated seismicity rates are likely to be synchronous with the global correlated seismicity rates along the global ridge system. Therefore, it leads that the global ridge activity is mainly governed by the seismicity activity of north of Iceland, Iceland, IOR, south of Australia, and Off Mexico regions. The kinematics of the long-term plate motion is basically controlled by the rotation poles of the plate but not by the plume motion beneath the hot spot on the ridge axis, so that the above-mentioned conclusion of major correlated seismic activity of the global ocean ridge may not be consistent with simple kinematic model

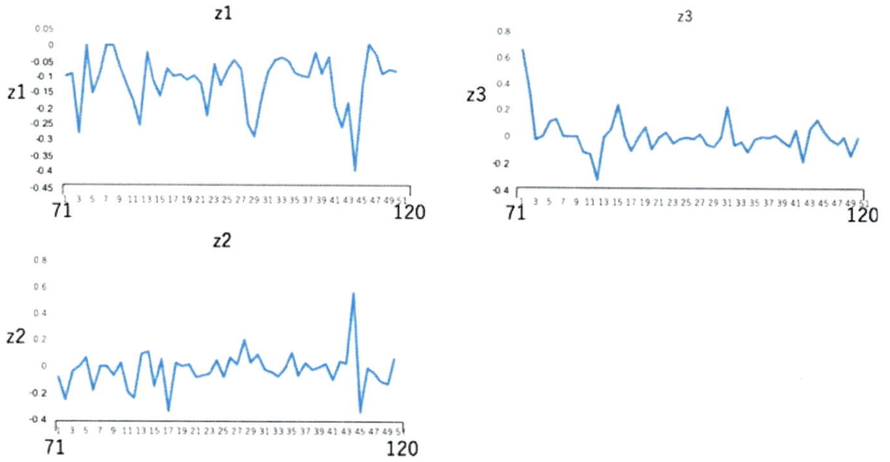

Fig. 5.49 Relations of the correlation intensity of the locality cells in the Indian Ocean Ridge and the correlated seismicity rates z_1–z_3. Horizontal axis is the number of locality cells in the Indian Ocean Ridge

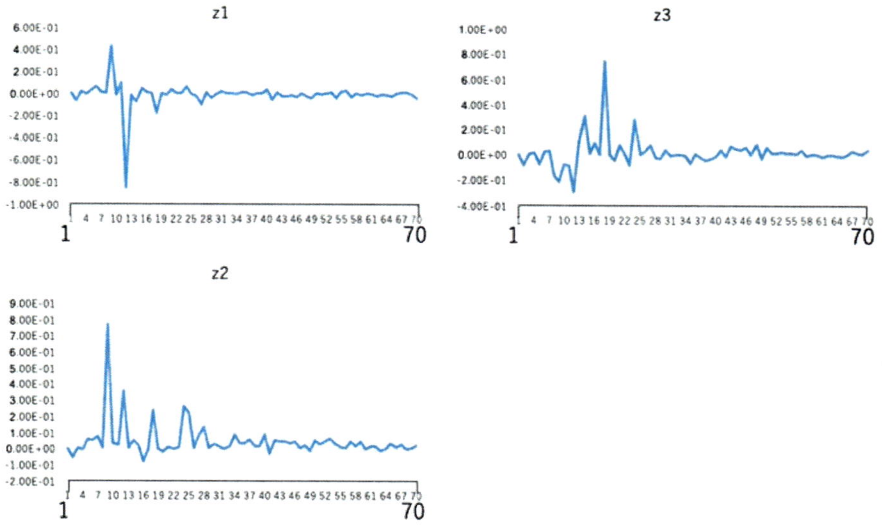

Fig. 5.50 Relations of the correlation intensity of the locality cells in the Mid-Atlantic Ridge and the correlated seismicity rates z_1–z_3. Horizontal axis is the number of locality cells in the Mid-Atlantic Ridge

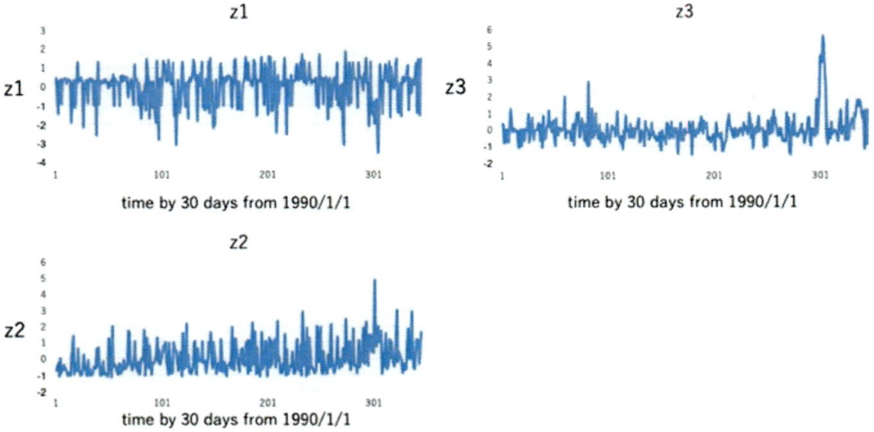

Fig. 5.51 Time series of the correlated seismicity rates z_1 to z_3 in the Mid-Atlantic Ridge from 1990 to 2018

of the long-term plate motion, but correlated seismicity rates of the spreading center may be governed really by the short-term plate motion in the several ten-year scales. It means that the plate motion in the short-term process is probably active but not passive by means of simple carpet model of the plate subduction. It then seems that the subduction process inferred by the seismicity activity of the global subduction boundaries is also controlled by the regional activity in the several local cells which are observed in the strongly correlated seismicity rates at NE Japan, TK, Chile, and Sumatra as seen in the previous chapters.

5.6 Global Seismic Activity of Subduction Zones and Oceanic Ridges

As seen in the times series of the global seismicity rates along the mid-oceanic ridges and subduction zones, we possibly compare the correlated seismicity rate fluctuations of them in the period from 1990 to 2018 for the seismicity of the magnitude 4 to 5 by the time window scale of 30 days. Figure 5.52 shows the comparison of correlated seismicity rate (z_i) time series of the global mid-oceanic ridge and subduction zones, being apparently similar fluctuation patterns between them. It seems that the correlated seismicity rate time series are characterized by the main trend in the z_1 and periodic variation in z_2 of the global subduction zone, though they are by gradual step-like increase of z_1 and weak periodic change of spike clusters of z_2 in global ridge zone.

The z_1 components of the global subduction zones and the mid-ocean ridges have similar trends as step-like change with each other, and therefore it seems that there

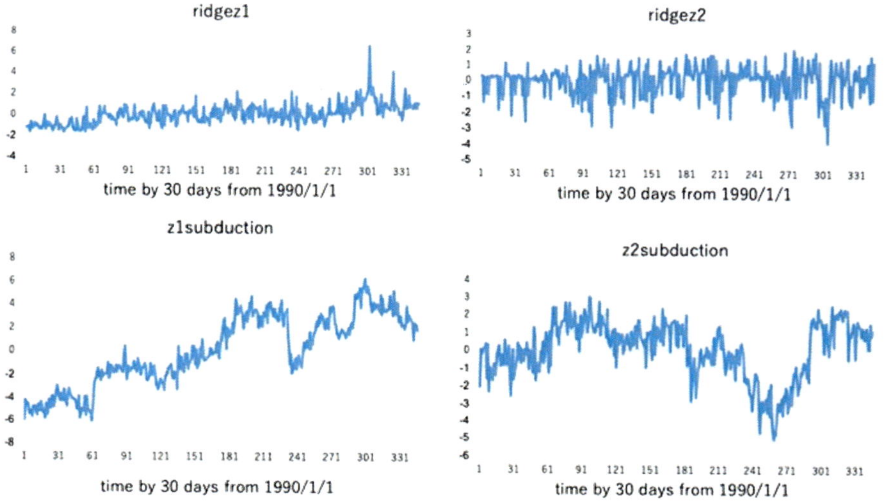

Fig. 5.52 Comparison of the correlated seismicity variations (z_1 and z_2) between global ridge (upper) and subduction zones (lower). Horizontal axis is the time by 30 days after 1990 to 2018

are three clusters in the diagram of z_1 of subduction zone and global ridge as shown in Fig. 5.52. On the other hand, the z_3 components of the global ridges show somewhat weak periodical patterns of spike clusters, although z_3 of the subduction zone shows the change from low to high level but there is the rapid increase and then recovery patterns between low and high z_3 level as shown in Fig. 5.52.

Therefore, in order to investigate the coherence of z_1 between the subduction zones and global ocean ridges, the phase map of z_1 between them is drawn in Fig. 5.53. It shows probably that the strong coherent activity of z_1 seismicity between them can be identified surely: that is, the z_1 strong correlated seismicity of the subduction zones increases with increasing those of the oceanic ridges with synchronous time series. The diagram (Fig. 5.53) that shows the relations between their both z_1 components clearly displays that the data points of the time increments make the distinguishable clusters with increasing the time from 1990 to 2018. This seems natural because the global plate motion should be satisfied with mass balance between the spreading and the subduction even if there are any relaxation time due to the elastic deformation of the plate itself within the time intervals (30 days) of the seismicity rate studied here.

It should be noted that the locality cells corresponding the z_1 components of the oceanic ridges are summarized to be Iceland, Galapagos, and Baja California and those of the subduction zones are to be NE Japan, Sumatra, Tonga-Kermadec, and Chile. And the global seismic moment release rates measured by the earthquakes over magnitude of 7.5 are also strongly correlated with the fluctuation patterns of z_1 noted previously (Fig. 5.38). It is, therefore, concluded that the correlated seismicity parameter z_1 changes coherently with the activity of the giant earthquakes which mainly contributes the global seismic moment release rate.

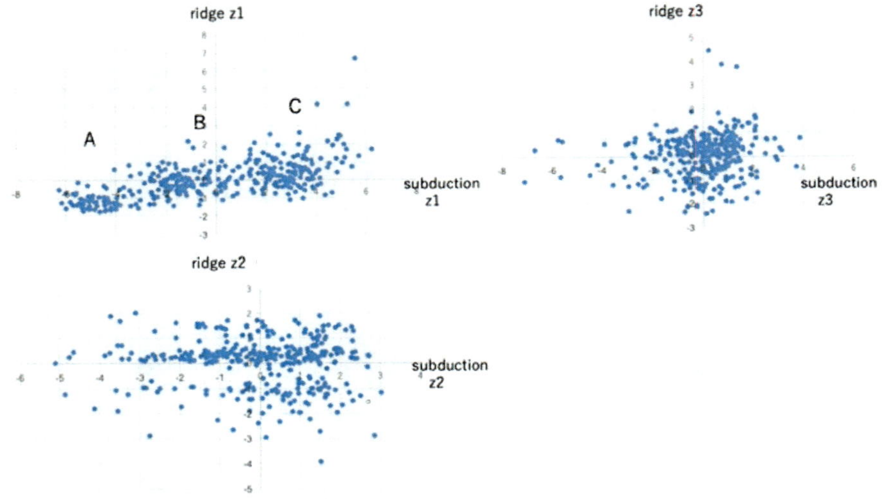

Fig. 5.53 Comparison of the correlated seismicity rate diagram of z_1 to z_3 between global subduction and oceanic ridge showing obviously three clusters A, B, and C with progressive order of the time

On the other hand, the z_2 components of the global ridges and the subduction zones are presented in the correlation diagram as shown in Fig. 5.53, indicating no coherent relations between them. Namely, it seems that the spike-like z_2 variation of the ridge seismicity is irrelevant with the long-term secular trend of the z_2 of the subduction zones seen in Fig. 5.53. Further, it is also noted that the z_3 variations of the ridges having very weak periodicity of about 3–4 years are apparently independent with the z_3 fluctuation of the subduction zones (Fig. 5.53).

References

Bishop CM (2006) Pattern recognition and machine learning. Springer, New York

Christensen K, Danon L, Scanlon T, Bak P (2002) Unified scaling law for earthquakes. PNAS 99(suppl 1):2509–2513. https://doi.org/10.1073/pnas.012581099

Ekstrom G (2007) Global seismicity: Results from systematic waveform analyses, 1976–2005. In Kanamori H (ed.) Treatise on geophysics. Elsevier, pp 473–481

Kanamori H, Anderson DL (1975) Theoretical basis of some empirical relations in seismology. Bull Seismol Soc Am 65, 1073–1095.e

Zapliapin I, Kreemer C (2017) Systematic fluctuations in the global seismic moment release, Geophys Res Lett 8. https://doi.org/10.1002/2017GL073504

Chapter 6
Correlated Seismicity of Japanese Regions

Abstract Seismic activity of the Japanese islands region is a part of the global seismicity dynamics, and then it should be governed by its boundary conditions embedded in the global seismicity dynamics. Thus, the downscaling of the global dynamics may be applied for the seismicity of the Japanese region.

Keywords Correlated seismicity time series of Japanese crust and mantle · Correlated seismicity patterns of partial b-value · Annual variations of the correlated seismicity

6.1 Outline of Tectonics of the Japanese Islands

The Japanese islands are the subduction plate boundary zones of the Pacific Ocean plate and Philippine Sea plate and Eurasian plate, and North American plate in the northern area. The seismic activity of these regions is very high, and the moment release rates are also large. The geometry of the subduction of the oceanic plate is different from northeastern islands (Hokkaido area) to southwestern islands (Okinawa and Yaeyama islands); in the Tohoku and Hokkaido regions, the subducted slab of the Pacific Ocean plate extends about 660 km depth and partly it makes stagnant slab in the transition zone of the mantle under the Eurasian plate (Fukao et al. 1992). On the other hand, along the Izu-Ogasawara-Mariana trench, the Pacific Ocean plate subducts under the Philippine Sea plate by the high angle and it extends in the lower mantle (ibid.).

Furthermore, along the southwestern Japanese islands (Honshu, Shikoku, and Kyushu), Nankai Trough subduction zone runs from Kanto to the Kyushu districts. The Philippine Sea plate subducts at the Nankai Trough with low subduction angle under the southwestern Japanese islands. The huge amounts of accretion complex

---

Electronic supplementary material The online version of this chapter (https://doi.org/10.1007/978-981-15-5109-3_6) contains supplementary material, which is available to authorized users.

M. Toriumi, *Global Seismicity Dynamics and Data-Driven Science*, Advances in Geological Science,
https://doi.org/10.1007/978-981-15-5109-3_6

consisting of Tertiary trench sediments involving some amounts of igneous intrusive masses such as Muroto gabbro mass form the fore-arc wedge body over the oceanic plate. The slab of the Philippine Sea plate subducts into the upper mantle beneath the San-in district of the Honshu and reaches about 40 km depth. Along the Nankai Trough, accretionary complex of the Cretaceous to Paleogene sedimentary rocks and the metamorphic rocks develops well from Kyushu to Kanto Mountains and partly it extends to the Ryukyu islands. On the other hand, the fore-arc wedge along the Ryukyu trench has not developed the sedimentary accretion masses developed along the Nankai Trough.

On the other hand, the Izu-Ogasawara-Mariana trench is the subduction boundary between the Pacific Ocean plate and the Philippine Sea plate and the slab of the Pacific Ocean plate subducts into the deep transition zone and also partly into the lower mantle region, being in contrast with the slab from the Japan Trench. The subduction angle of the slab reaches about 90° but near the trench it does about 10–30°. It is noted that there are the arrays of serpentine seamounts which is called as conical seamount, and it is considered that the source of serpentine is the hydrated wedge mantle. The serpentine rocks of this seamount are brecciated and powdered, and cemented by carbonate matrix, and occur as the mudflow and debris flow. This seems to be called as the serpentine mud volcano (Fryer et al. 1985).

The back-arc opening occurred behind the Japanese islands during the periods from 30 to 10 Ma, and it formed the Sea of Japan. The remnants of the back-arc opening are the Yamato bank but the oceanic crust formed by the spreading seems to be thick (~20 km) rather than those near the mid-oceanic ridges having about 5 km thick. Along the Ryukyu trench the back arc opening makes the rift zone named as the Ryukyu basin, though the igneous activity is not basaltic but dacitic to rhyolitic.

The volcanic activity along the northeastern Japan arc is characterized by regularly alignment patterns of volcanos which are called as the volcanic front by Sugimura (1960) and Tatsumi (1986). The volcanos behind the volcanic front are also aligned but a little irregularly. On the other hand, in the southwestern Japan arc there are a few active volcanoes arrays parallel to the Nankai Trough from Chugoku to south of Kanazawa, but along the Ryukyu islands to Kyushu active volcanos and submarine volcanos occur as the clear volcanic front. The Aso Caldera is the typical large Caldera forming active volcano in the central Kyushu, and it located at the center of the east-west running rift zone.

The Izu-Ogasawara-Mariana arc contains two islands arrays composing non-volcanic islands and volcanic islands. The Izu-Bonin-Mariana arc is called as IBM volcanic arc. In the crossing regions of Northeast Japan volcanic arc and IBM arc, Mt. Fuji and Hakone Caldera locate. The islands array of Ogasawara is called as the outer non-volcanic arc, and it is associated with the fore-arc basin.

6.2 Seismicity of Japanese Islands Region

Figure 6.1 displays the distribution patterns of earthquakes hypocenters with magnitudes from 2.0 to 3.0 during 1998 and 2015 plotted by using the cloud software "TSEIS" in the ERI data center. The distribution patterns of the earthquakes with hypocenters above 30 km depth show the dense feature extended in the whole Japanese islands arc and fore-arc regions. To view the more detailed features of the high-density zones of the earthquakes hypocenters in the shallower layers of the crust and uppermost mantle, the distribution maps in the periods of 2001/8/1-9/1 and 2016/8/1-9/1 are presented in Fig. 6.1. It is obvious that four arrays of seismic active zone can be assigned nearly parallel to the axis line of the Japan Trench in the Hokkaido to Kinki districts, and two arrays are possibly seen to be parallel to the Ryukyu trench. The depth of these alignments is ranged from 0 to 20–30 km as shown in Fig. 6.2. The former period corresponds to the relative inactive time and the latter does to the active time judging from the total numbers of the small earthquakes during the one month: The number of August in 2001 is 1814 but that of August in 2016 is 3503. These dense alignment structures (called as seismic lineament hereafter) of the hypocenters of earthquakes suggest reasons whether these abnormal seismic lineaments are derived from the stress concentration or from the strain concentration. Recent investigations by Iio et al. (2018) have clarified that the differential stress along the abnormal seismic lineaments estimated from the stress drop at the large earthquakes is small enough rather than that of their outer zones. Therefore, it seems to be concluded that the abnormal seismic lineaments are the seismic weak zone and also are the active deformation zone with brittle and plastic strains of constituent crustal rocks.

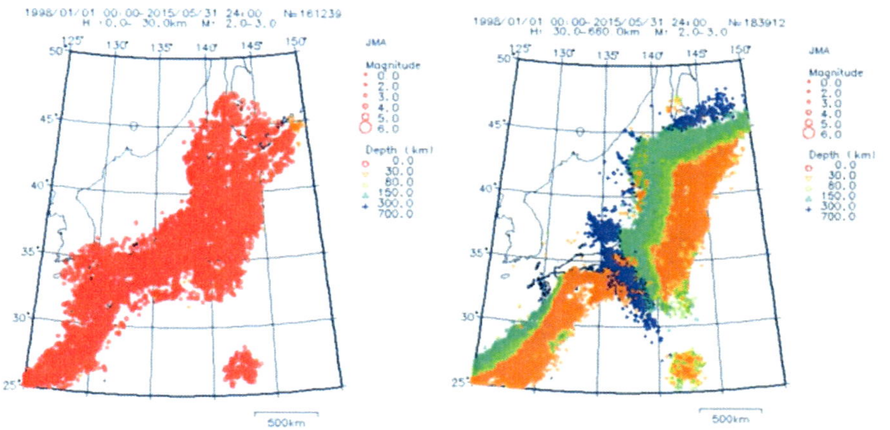

Fig. 6.1 Distribution map of hypocenters of earthquakes (M2–M3) in the crust (left; 0–30 km depth) and mantle (right; 30–660 km) in the Japanese islands region from 1998 to 2015 made by "TSEIS" of ERI, University of Tokyo

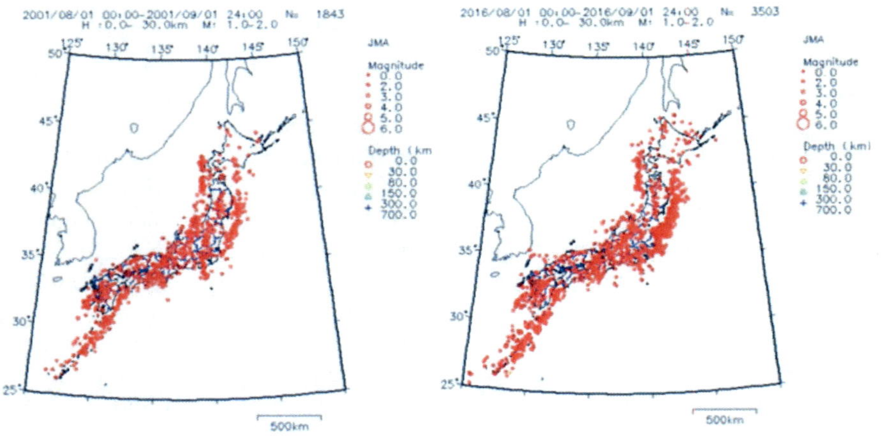

Fig. 6.2 Comparative distribution maps of earthquakes hypocenters (M1–M2) of inactive (left; 2001) and active (right; 2016) periods in the Japanese islands region, showing the zonal distribution of dense hypocenters

The active deformation zones also mean the large strain zones that are probably composed of cataclastic altered rocks formed by extensively reacted with high-temperature solutions. And the high-temperature fluid seems to be derived from circulated fluid together with fluid from dehydrated slab. In NE Japan arc, the abnormal lineament seems to lie along the axial zone of the volcanic arc, and the chemistries of the magma along this zone appear the signatures from the slab rocks as studied by Kimura (2017).

In SW Japan arc, it is recognized that there are two abnormal seismic lineaments from the Kyushu to Ryukyu islands as shown in Fig. 6.2, and the fore-arc lineament runs at the Bungo straight between Shikoku and Kyushu where the slow slip events have occurred, being sandwiched by abundant non-volcanic tremors zones in the regions of its north and south. Furthermore, it should be noted that the northern lineament in the central Kyushu district is just overlapped in the area of the 2016 intraplate large earthquakes with M6.7 and the extensional zone of the large Caldera of Aso volcano.

In the mantle zone of the Japanese islands regions, there are several abnormal seismic lineaments behind the trench axis: In the periods between 2000–2003 and 2014–2016, the highly active lineaments are distinctive parallel to the axes of Japan Trench and Ryukyu trench as shown in Fig. 6.3. On the other hand, there appears the abnormal seismic lineament just near the trench axis in the period from 2014 to 2016 after the giant Tohoku-Oki earthquake. The intermediate and deep focused earthquakes also display the seismic lineaments along the subducted slab depth contours of 80–100 km and 400–500 km depth.

The seismic layer named as the Wadachi–Benioff zone beneath the islands arc of the Japan becomes deepen behind the trench in the Japan Trench, Izu-Mariana trench and Ryukyu trench as described earlier. The lineaments appeared in the seismic

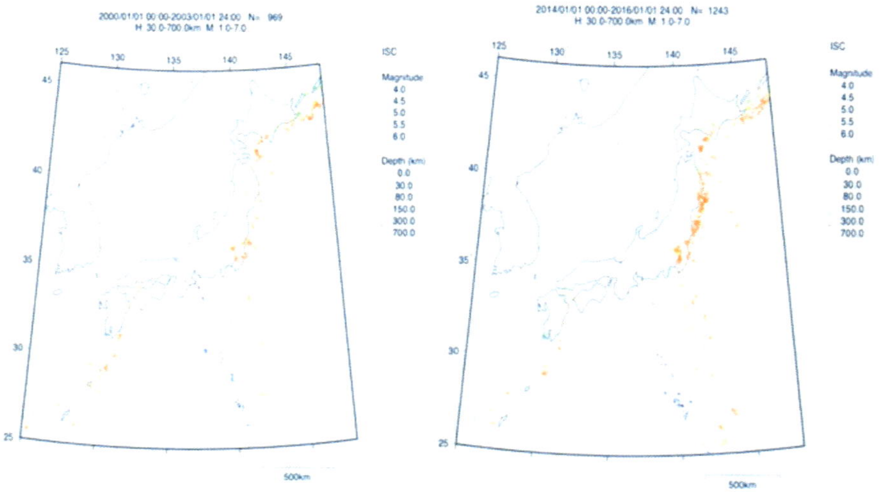

Fig. 6.3 Comparison of distribution patterns of earthquake hypocenters (M1–M7) in the inactive (left; 2000–2003) and active (right; 2014–2016) periods in the mantle under the Japanese islands region, showing the dense distribution zone along the trench

maps mentioned above seem to manifest the existence of seismic clusters along the subducted slab of the Pacific plate and Philippine Sea plate at the regions of 30–100 km and 400–500 km depth. The former clusters seem to be due to fracturing resulted from dehydration from antigorite to olivine plus H_2O as suggested by Omori et al. (2002).

The non-volcanic (NV) tremors that are recently found as the swarm of small and low-frequency earthquakes occurred along on the subducting slab, and these are highly correlated with the slow slip event (SSE) in Shikoku to Kii peninsula (Obara 2002; Shelly et al. 2007). The northern alignment of NV tremors runs almost always at the depth near the bottom of the arc crust along the Philippine Sea plate but the southern one does shallower boundary of the Philippine Sea plate under the south of Bungo straight. The boundary region between these tremor regions comprises of slow slip event zone. These slip activities are mutually but weakly correlated with each other, although in the case of Cascadia subduction zone of west of Vancouver islands the clear coherency between SSE and NV tremor activity. As stated in the previous chapters, there is a difference in the fracture mechanism among SSE, NV tremor and regular earthquakes occurred in the overlapping regions of the subduction boundary in terms of the relationship between the magnitude and the duration of seismicity (Ide and Shelly 2007). The characteristic slip feature of them is probably responsible for the shear slip instability having special timescales from incubation slip to cessation of slip of the boundary interface. However, it is very difficult to be explained that the range of timescales of SSE is very wide from several minutes to several years, but it suggests the coupling of water migration and shear crack propagation.

6.3 Seismicity Cloud of Japanese Islands Crust and Mantle

The studied cell locations for data-driven science method described in the previous chapters are shown in Fig. 6.4, and these are stored in the database of ESM. The cell locations are assigned as the number from 1 to 126. The cell size is defined as the width of 5° of longitude and as the length of 2.5 degree of latitude and depth of 0–30 km for the crust and 30–700 km for the mantle. Thus, the area correction factors are given by the 1/(cos (π L/180)) as same as the case of global seismicity study.

The seismic events in the ranges from magnitude 1 to 2, 2 to 3 and 3 to 4 are counted by sorting the event numbers of each cell from earthquakes list in the database of JMA-1 of ERI and NIED in the period from 1998/1/1 to 2018/8/29. Total seismic events to be sorted are over 800,000 using my own made software in R language. It takes about 15 min to perform sorting of the data in every magnitude screening above. In this study, the newly made database of these event numbers are counted in the time windows of 1 day, 3 days, 5 days, 10 days, 30 days, and 50 days for validation of different time window processing of seismicity rate data. Of course, the database of the 1 day time window is the basic one.

The raw seismicity rate data defined above are shown as the time series of many constituent vectors in the Japanese islands region from n_1 to n_{126} where n_k is the seismicity rate time series of the kth cell. Therefore, n_k here contains m samples of the time series and so it is defined as the data matrix from the time series as,

$$N = \{n_{ij}\} \quad (i = 1, 2, \ldots, m; \; j = 1, 2, \ldots, 126) \tag{6.1}$$

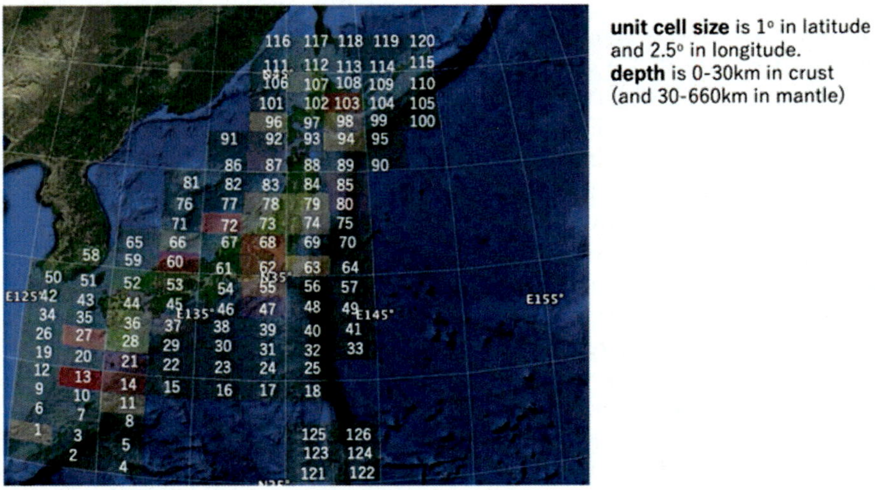

Fig. 6.4 Location and locality cell number of the Japanese islands region in the Google Earth map to investigate the correlated seismicity rate by data-driven science

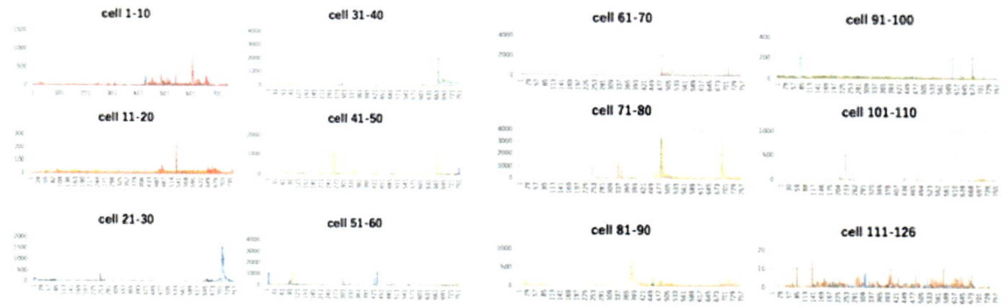

Fig. 6.5 Left, time series of the seismicity rates n of the locality cells in the Japanese islands region from 1998 to 2018. The horizontal axis is the time by 10 days and vertical axis is the number of seismic events (M1–M2) per cell and 10 days

where $m = 7540$ for 1 day window.

The time series of $n_j(t)$ in the case of magnitudes of M1–M2 is displayed in Fig. 6.5, suggesting the strongly noisy patterns with several peaks during the studied time range. There are large peaks at about 2000, 2011, and 2016 corresponding the large earthquakes in Japanese islands, that is, Tottori earthquake, Tohoku-Oki earthquake, and Kumamoto earthquake, respectively. There is weak activity of seismicity rates in northern Hokkaido and Ogasawara cells. The general trends of seismicity rates in almost always cells show the low activity before 2004 and high activity after 2004 in the crustal layers. On the other hand, the seismicity rate time series of the mantle which can be found in Fig. 6.8 appear somewhat random patterns as like as those of the crust noted above. However, there seem to be several peaks near 2000–2005 and 2011–2016. It is difficult to conclude that there are systematic variations in the time series of various cells of the Japanese islands region.

To illustrate the seismicity rate data cloud in the high-dimensional space of the cells, first let us transform the seismicity rate data to the logarithmic values as described in the previous sections. The transformed logarithmic seismicity rate matrix is composed of

$$x_{ij} = \log\left(1 + n_{ij}\right) \quad (i = 1, 2, \ldots, 754, \; j = 1, 2, \ldots, 126) \tag{6.2}$$

In this section, the comparison of the data cloud between the crust and mantle of the Japanese regions is as follows. The data cloud of the crust is shown in the x_{10}, $x_{62,}$ and x_{68} three-dimensional diagram (Fig. 6.6). On the other hand, that of the mantle is shown in the same Fig. 6.7 as well in the x_{11}, $x_{64,}$ and x_{70} diagram. Both data clouds appear typical patterns that show central spherical dense cloud and several spines extending from the central cloud. The mean value of total data clouds locates at the position a little apart from the center of the central cloud. The mean point of the data cloud is referred in the preceding sections to discuss the correlated seismicity rate time variations.

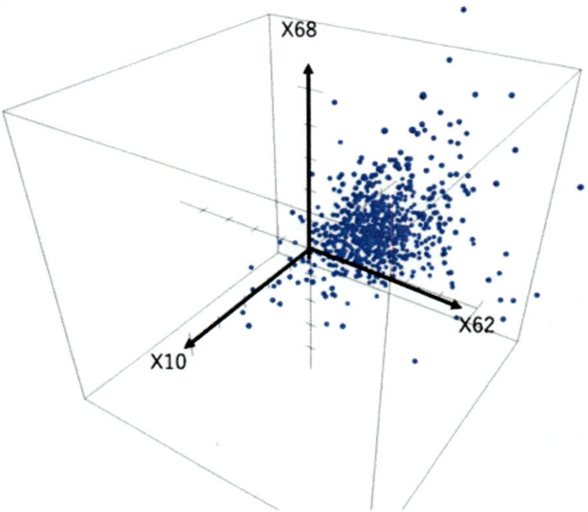

Fig. 6.6 Seismicity cloud by logarithmic seismicity rates x of representative locality cells in the crust of Japanese islands region from 1998 to 2018

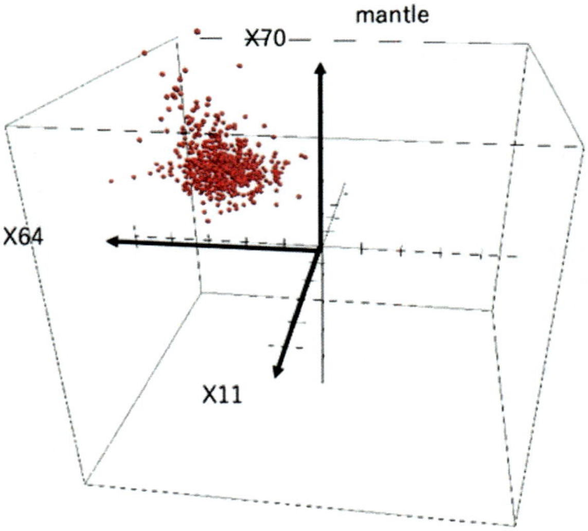

Fig. 6.7 Seismicity cloud by logarithmic seismicity rates x of representative locality cells in the mantle of Japanese islands region from 1998 to 2017

The seismicity data cloud is basically expressed in the high-dimensional space, and in the cases of Japanese crust and mantle it should be appeared in the 126-dimensional space. To discuss the characteristic features of the high-dimensional seismicity cloud, the newly orthogonal basis vector coordinate system should be introduced using the dimension reduction methods. In these figures (Fig. 6.8) illustrating the characteristic seismicity cloud even in the three-dimensional space, there are important features showing the spherical dense core and spines that indicate the extension from the core toward the composite direction of several axes (locality

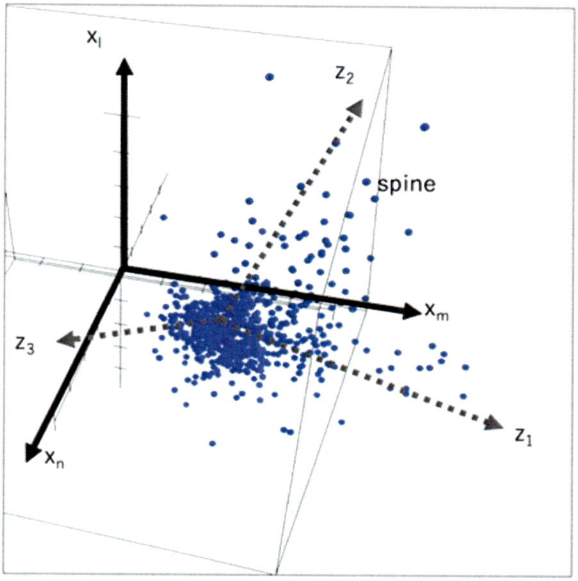

Fig. 6.8 Seismicity cloud by logarithmic seismicity rates x of representative locality cells and the schematic illustration of the correlated seismicity rates. Scattered plots from dense core are called as spine in this book

cells). It should be noted that the spines, therefore, mean the correlated seismicity of several cells in the studied regions, suggesting the coherency of seismic activity of different cells both in the regional and global scales.

As discussed in the previous sections, it is critically interesting that there is an invariant or nearly invariant set of the seismicity cloud as shown in Figs. 6.6 and 6.7. If it is possible to assume, the seismicity cloud should be complete for reconstruction of the dynamic model in the global and regional seismic activity. Considering that there are recurrence times over several hundred years of giant earthquakes in regional and global scales, many other spines which still do not appear in the diagram mentioned above will be expected undoubtedly. Regarding the case of Japanese islands regions, there occurred many giant earthquakes exceeding M8 after AD1000. Thus, the seismicity cloud studied here in the period from 1998 to 2018 is still not enough to reconstruct an invariant cloud, although as discussed later the enough periodicity of correlated seismicity rates can be recognized from the available dimension reduction methods. Therefore, it seems that the several spines appeared in the seismicity cloud are potentially discussed the systematic variations of each spines and their dynamics.

6.4 Characteristic Features of the Correlated Seismicity Rates

In order to investigate the correlation mode of the seismicity rates in the regional scales of the studied areas, the coordinate transformation of logarithmic seismicity

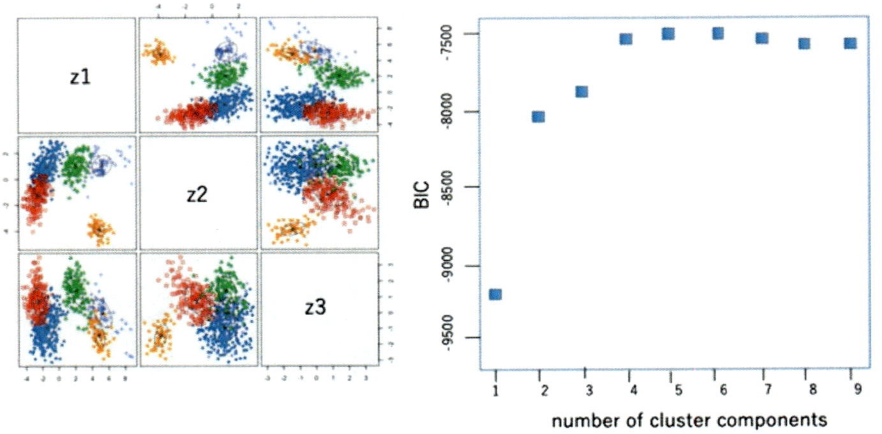

Fig. 6.9 Clustering of the correlated seismicity rates z1 to z3 of the crust in the Japanese islands region from 1998 to 2018 by means of "mclust" package in R and BIC for the clustering

rates has been carried out by means of rotation method by eigenvectors of their covariance matrix as defined by deterministic PCA and stochastic PCA (sparse PCA) method. The comparison between these two methods is tried to validate and verify the results.

The seismicity data matrix is constructed for the following conditions: The time windows are 3 days, 5 days, 10 days, 15 days, 30 days, and 50 days, and space windows are cells of 2.5° of latitude and 5 degree of longitude, and of 5 degree of latitude and 5° of longitude. The comparison of data processing is made by the Bayesian information criterion (BIC; Bishop 2006) for stochastic clustering of correlated seismicity rates in the reduced principal subspace as shown in Fig. 6.9. For the BIC value is larger, the more verifying the data processing is, it seems that the validation of the time windows is justified for larger value but the large values of the time window mean the averaging of the time series of seismicity rates fluctuation. Therefore, the validation of the data processing may be evaluated by the robustness of BIC against the change of sample number as shown in the bottom of Fig. 6.10. Accordingly, the most robust sampling by changing the sample number of the time series is satisfied in the time window of 10 days.

Furthermore, the comparison between the deterministic PCA and stochastic sparse PCA has been carried out for validation for characteristic features of correlated seismicity rates variations as shown in Figs. 6.11 and 6.12. There is an enough similarity between them, indicating the sparse PCA method is also available for the investigation of the correlated seismicity rates variations. It is also clear that the BIC of the clustering model by deterministic PCA and that by sparse PCA is not different so much and that both methods are equivalently available for the investigation of correlated patterns of seismicity.

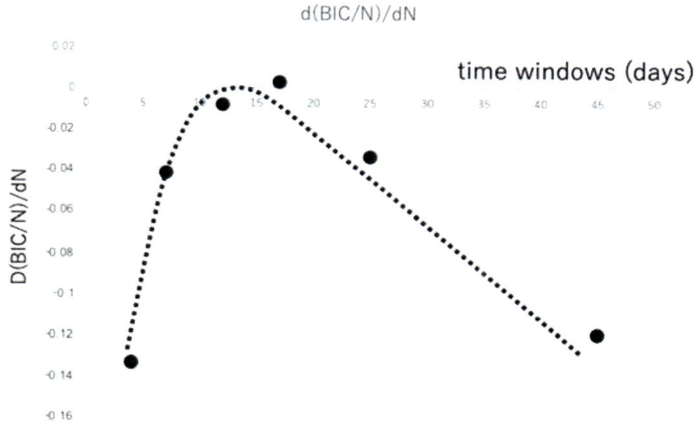

Fig. 6.10 Relation of the rate of Bayesian information criterion (BIC) by number of samples (N) and the time windows (horizontal axis) in the correlated seismicity rates of the Japanese islands region

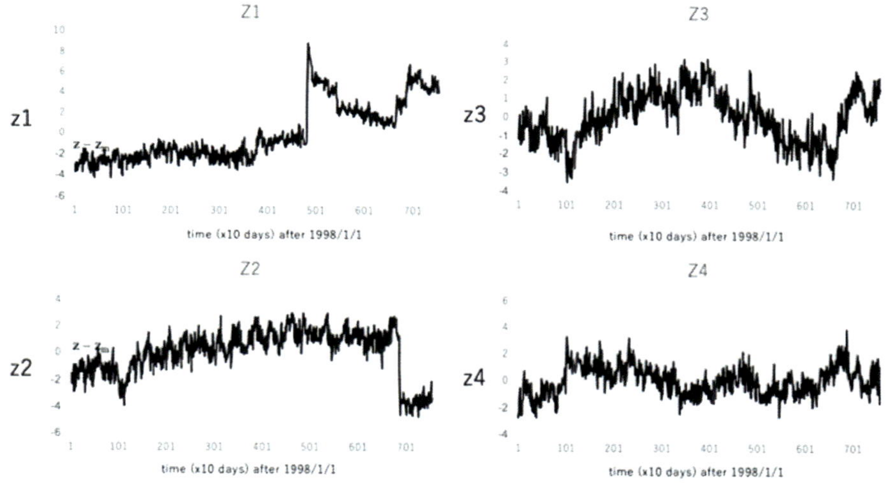

Fig. 6.11 Time series of the correlated seismicity rates z_1 to z_4 (for M1–M2) of the crust in the Japanese islands region from 1998/1/1 to 2018/8/28

6.5 Characteristic Features of Correlated Seismicity Rate Time Series

The data transformed by means of translation and rotation methods using the eigenvectors of covariance matrix of logarithmic seismicity rates are shown as the time series of z_1 to z_4 (Figs. 6.11 and 6.12), and those of the higher order of z_i ($i > 5$) display

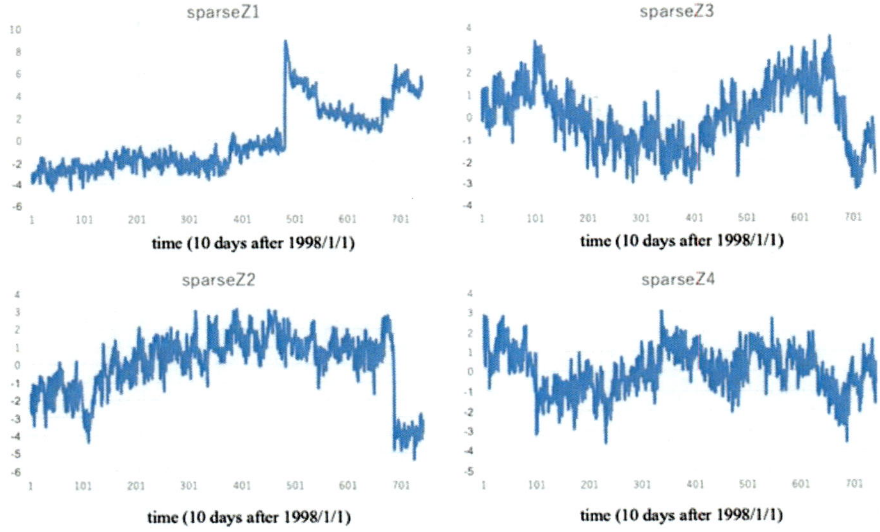

Fig. 6.12 Time series of the correlated seismicity rates z_1 to z_4 (for M1–M2) by sparse PCA of the crust in the Japanese islands region from 1998/1/1 to 2018/8/28

the Gaussian noisy patterns as shown in Fig. 6.13. Therefore, it seems probable that the characteristic features of logarithmic seismicity rates matrix are represented mainly by z_1 to z_4 in the case of deterministic PCA. In order to check the validity of the efficient numbers of parameter, the dimension reduction method by sparse PCA (with LASSO) was carried out for above data matrix as shown in Fig. 6.12.

The characteristic feature of the z_1 time series of M1 to M2 for the crust (depth range from 0 to 30 km) displays the stepwise increase from 1998 to 2011, and then it abruptly increases after 3.11 of 2011 of Tohoku-Oki earthquake. After 3.11, the z_1 shows a gradual decrease and again repeating up and down until 2018. It seems that the change after 3.11 of 2011 during several years is responsible for the post-seismic activity of the Tohoku-Oki earthquake, and up and down of the z_1 after 2016 is probably due to large intraplate earthquakes at Kumamoto and Fukushima. The time series of the z_2 component, on the other hand, displays the half periodic pattern

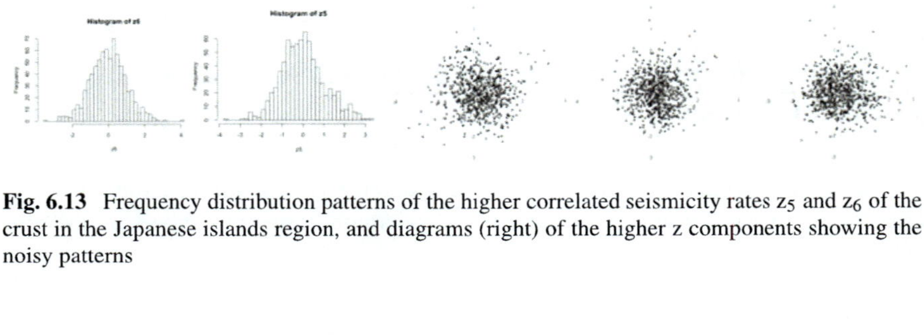

Fig. 6.13 Frequency distribution patterns of the higher correlated seismicity rates z_5 and z_6 of the crust in the Japanese islands region, and diagrams (right) of the higher z components showing the noisy patterns

with sudden decrease at 2016, as being later than the Kumamoto earthquake. This rapid change of the z_2 cannot be clearly assigned by any large earthquake of Japanese islands region, and further, it disappears in the z_2 pattern of the NE Japan regions as studied in the later sections, thereby suggesting that the abrupt change of the z_2 may be responsible for the seismicity change in SW Japan regions. The small change in the z_2 in 2000 is due to the intraplate Tottori earthquake and the small increase of the z_2 appeared at 2016 before the rapid decrease clearly coincides with the Kumamoto earthquake.

The z_3 and the z_4 patterns indicate obviously the cyclic ones: The z_3 time series shows two peaks during 1998 to 2018, suggesting the periodicity around 15 years, but the z_4 one does three peaks and periodicity around 5–8 years. The periodicity of them does not seem synchronizing with each other. The peaks of the z_3 are in 2000 corresponding to the Tottori earthquake and in 2015 before the Kumamoto earthquake.

The large step-up at 2011/3/11 of the Tohoku-Oki earthquake may be eliminated by means of the Omori equation (or ETAS model; Ogata 1988) applicable for the post-seismic activity using equation of $N(t) = N_o/(t + c)$ which N_o is the number of micro-earthquakes at the time of Tohoku-Oki earthquake and t is the time after that earthquake. In this case, the characteristic features of z_1 and z_2 are shown in Fig. 6.14. It seems obvious that the feature patterns are similar to those in the previous z_1 and z_2 without recalculation by the Omori equation except the large jump and relaxation after the Tohoku-Oki earthquake. Thus, in order to discuss the micro-earthquake

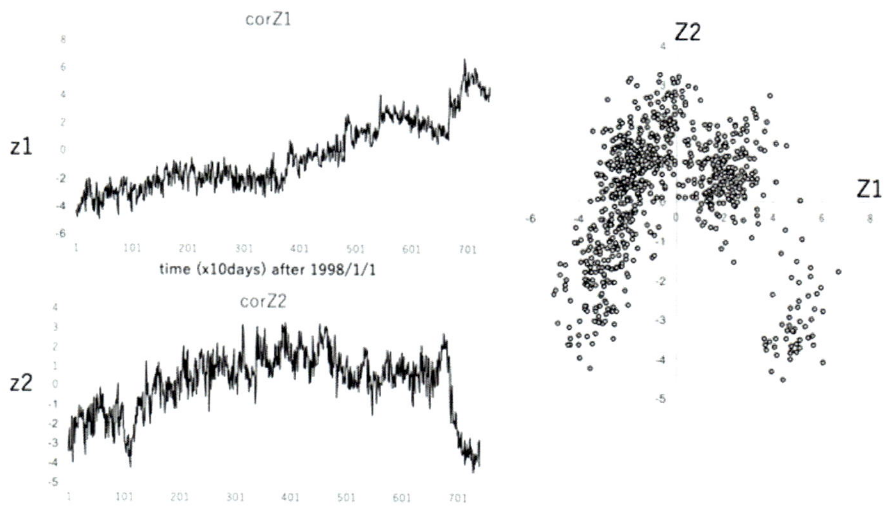

Fig. 6.14 Time series of the correlated seismicity rates modified by Omori equation (see text) and the z_1–z_2 diagram of the crust in the Japanese islands region

activity of the periods before and after large earthquakes and co-seismic activity, it needs the z_1 and z_2 variations of seismicity rates without recalculation by the Omori equation.

As stated previously, the z_i time series obtained by sparse PCA method (with LASSO) is indicated in Fig. 6.12, thereby showing that there is a clear similarity in the characteristic features of the z_i time series between the deterministic PCA and probabilistic sparse PCA methods. The cross-validation in the sense of comparative characteristic features of z_i therefore suggests that both methods are available for the extraction of the features of seismicity rate time series.

The weight parameters of the eigenvectors obtained by large eigenvalues of covariance matrix of seismicity rates data can be expressed by their intensities in the series of cell number meaning the labels of the locality cells of Japanese islands region (Fig. 6.15). Here, the weight parameters of the eigenvectors are to be the basis vector components in the feature space z and means the components of orthogonal rotation matrix. Therefore, the diagram showing this weight strength versus locality cell number indicates the characteristic correlation of the seismicity rates of the present cells. Then, the strongly correlated seismicity rates cells can be easily found by this type of diagram for Japanese islands crust (Fig. 6.15).

In Fig. 6.16, the schematic diagram of strongly correlated cell numbers is shown in the weight intensity diagrams of z_1, z_2, z_3, and z_4 in the crust of Japanese islands regions. It is obvious that there are strong correlated cells both of Okinawa to south of Kyushu and the plate boundary zone of northeastern Japan. On the other hand, the map showing the weight coefficient intensity of the basis vectors displays the strong correlated seismicity rates in the locality cells of Okinawa, west of Kyushu, Tottori,

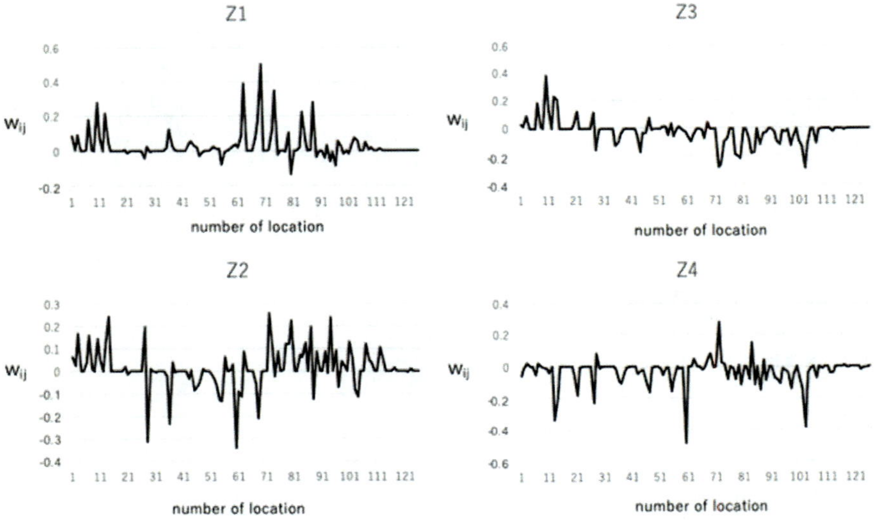

Fig. 6.15 Linear patterns of the correlation intensity (w_{ij}) of the locality cells shown in Fig. 6.4 of the crust in the Japanese islands region

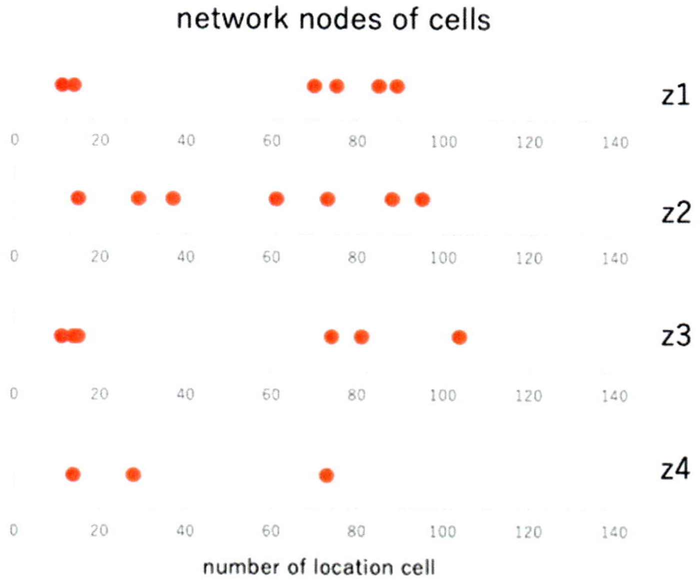

Fig. 6.16 Representative locality cells (network nodes of the cells) of the correlation intensity ($|w_{ij}| > 0.2$) for z_1 to z_4 of the crust in the Japanese islands region

Kanto, and Tohoku regions (Figs. 6.16 and 6.17). In this map of the z_3, there is strong contribution of the Okinawa seismicity correlated with NE Japan regions. It is noted that the strength of cells sometimes shows negative value but it is responsible for the new basis vector direction such as z_2 involving the negative direction of original cell basis vectors. Thus, even if the weight coefficients in new basis vectors z_i are of negative values, the absolute value of them are indicative of the strength of the correlation of seismicity rates of the cells.

The time series of z_1, z_2, z_3, and z_4 for the correlated seismicity rates of the mantle in the Japanese islands region are shown in Fig. 6.18. The characteristic features of them are similar to those of the crust as shown earlier: The feature of z_1 time series displays the long-term secular trend and the periodical change with two peaks, and that of the z_2 shows the cyclic patterns with two peaks. On the other hand, the patterns of z_3 and z_4 display a little cyclic change with large amplitude of noisy variations. As stated later, it is noteworthy that there are somewhat distinctive annual variations in the time series of the z_1. Furthermore, it seems to be important that there are no clear signals of giant plate boundary earthquake of the Tohoku-Oki 2011 both in the z_1 and z_2 time series, but that only in the z_3 time series a small jump at 2011/3/11 can be identified.

The strength of correlated seismicity rates of the mantle under the Japanese islands regions can be seen in the weight parameters intensity diagram as mentioned above (Fig. 6.19). The patterns of weight parameters intensity reveal that the z_1 pattern is composed of many correlated locality cells among Okinawa, Kyushu, Tottori, Kanto,

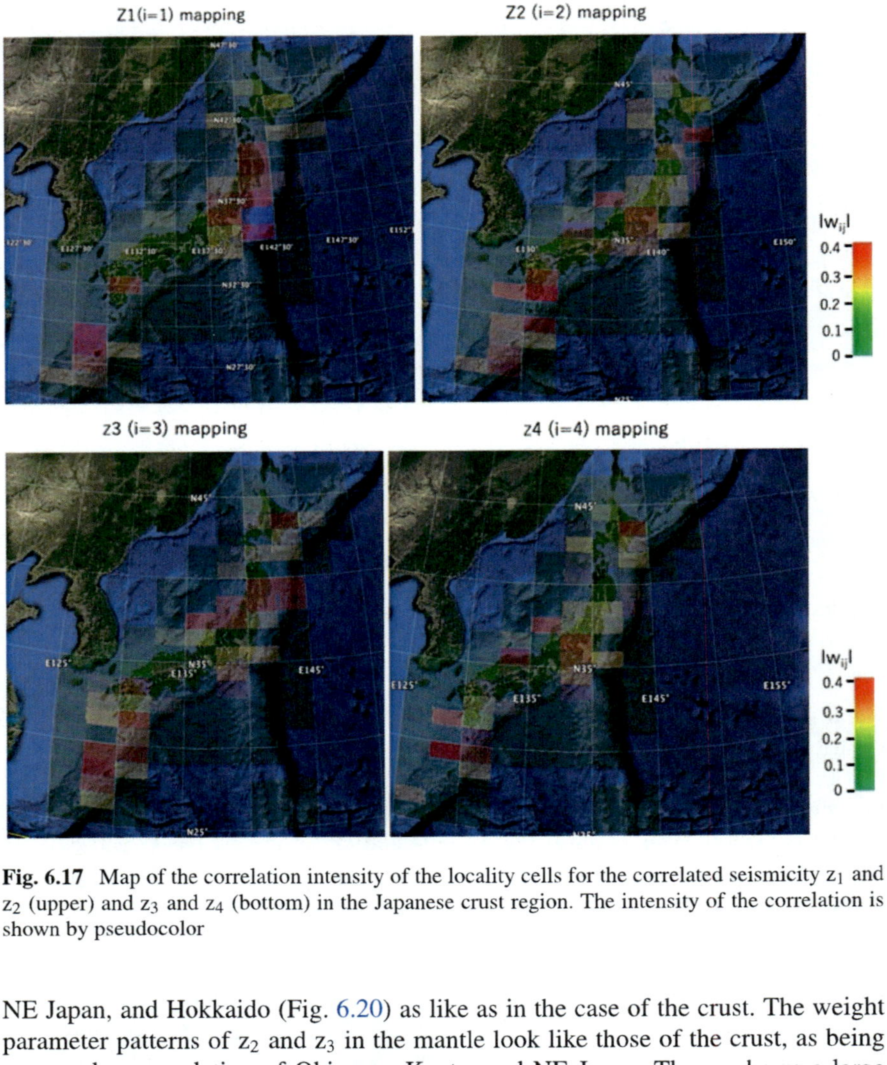

Fig. 6.17 Map of the correlation intensity of the locality cells for the correlated seismicity z_1 and z_2 (upper) and z_3 and z_4 (bottom) in the Japanese crust region. The intensity of the correlation is shown by pseudocolor

NE Japan, and Hokkaido (Fig. 6.20) as like as in the case of the crust. The weight parameter patterns of z_2 and z_3 in the mantle look like those of the crust, as being appeared as correlation of Okinawa, Kanto, and NE Japan. The z_4 shows a large contribution of NE Japan.

Next, let us discuss the characteristic patterns of the features time series of z_i of the seismic magnitude 2–3 levels as shown in Fig. 6.21. The similar patterns of z_i are seen in Fig. 6.21, thereby indicating these of the z_i for magnitudes of 1–2 as shown previously. The feature pattern z_1 of magnitude 2–3 displays a slow decrease from 1998 to 2011 and rapid increase at 2011/3/11, and then it shows the relaxation decrease for several years but keeps at high levels during 2013–2018. On the other hand, the pattern of z_2 displays also a similar variation as that of z_1 before 2011, and it abruptly decreases and slowly recovers; but after 2016, it again increases to high

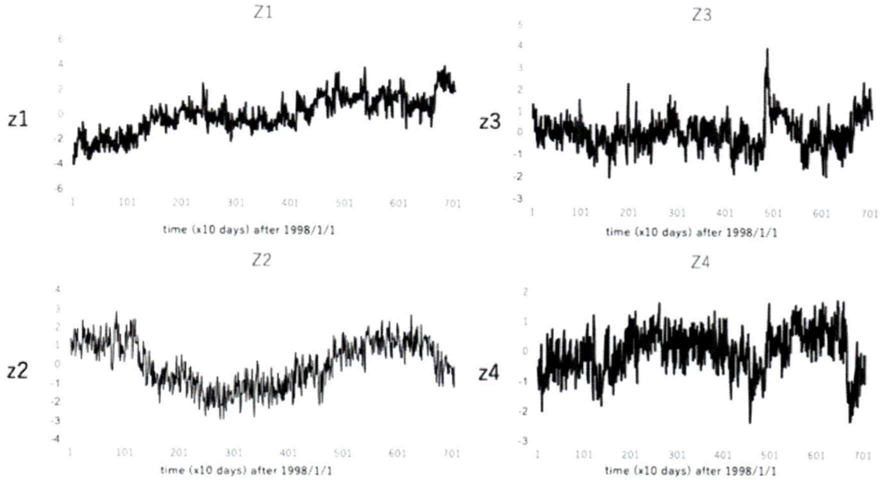

Fig. 6.18 Time series of the correlated seismicity rates z_1 to z_4 (for M1–M2) of the mantle in the Japanese islands region from 1998/1/1 to 2017/4/30

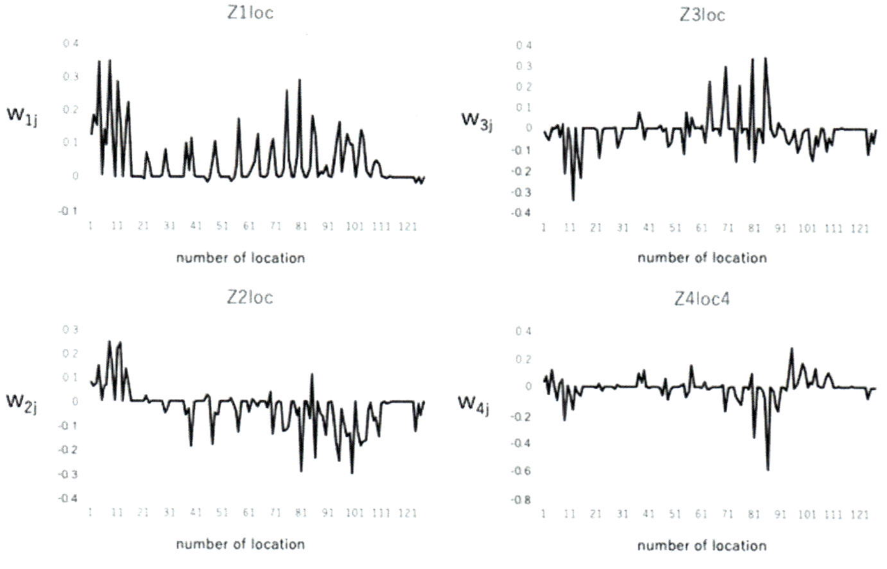

Fig. 6.19 Linear patterns of the correlation intensity (w_{ij}) of the locality cells shown in Fig. 6.4 of the mantle in the Japanese islands region

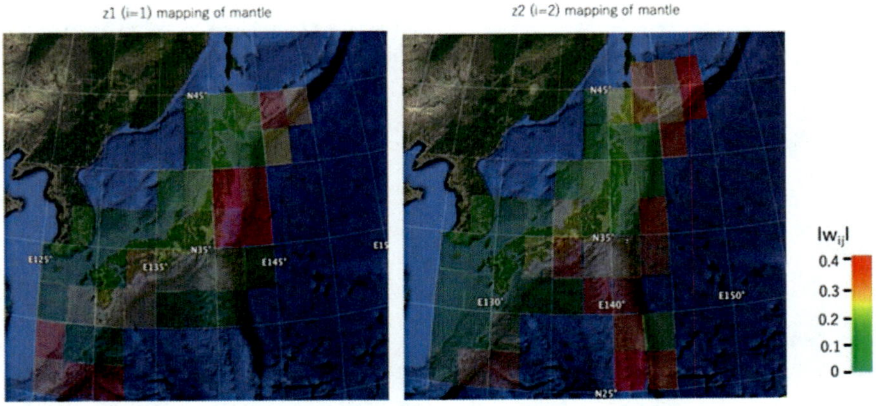

Fig. 6.20 Map of the correlation intensity of the locality cells for the correlated seismicity z_1 and z_2 in the Japanese mantle region. The intensity of the correlation is shown by pseudocolor

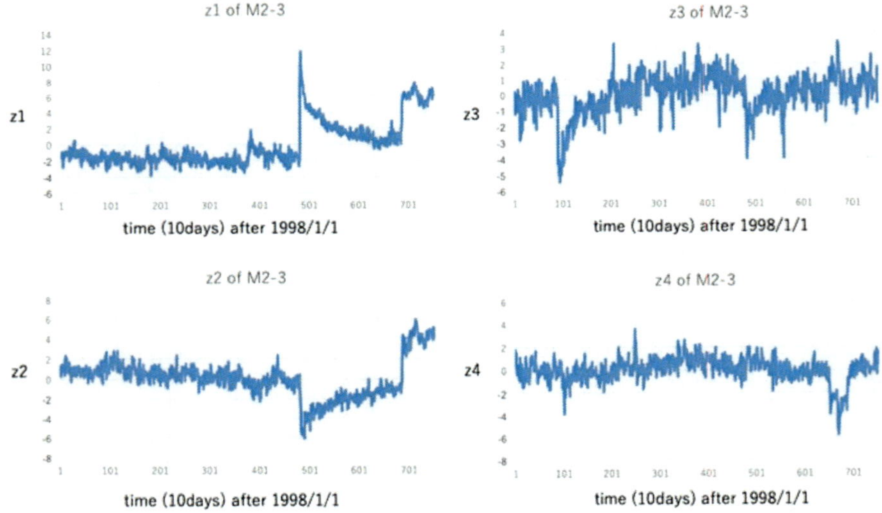

Fig. 6.21 Time series of the correlated seismicity rates z_1 to z_4 (for M2–M3) of the crust in the Japanese islands region from 1998/1/1 to 2018/8/28

level. The intensity of the correlation of seismicity rates in Japanese islands region is characterized by the high contribution of the NE Japan in z_1 and wide regions of high intensity of correlated seismicity rates as shown in Fig. 6.22. Judging from the modified variation pattern by Omori equation as noted previously, the pattern during the period from 2011 to 2016 is not wholly derived from the after seismic micro-earthquakes but from the correlated general seismicity trends of the Japanese islands regions.

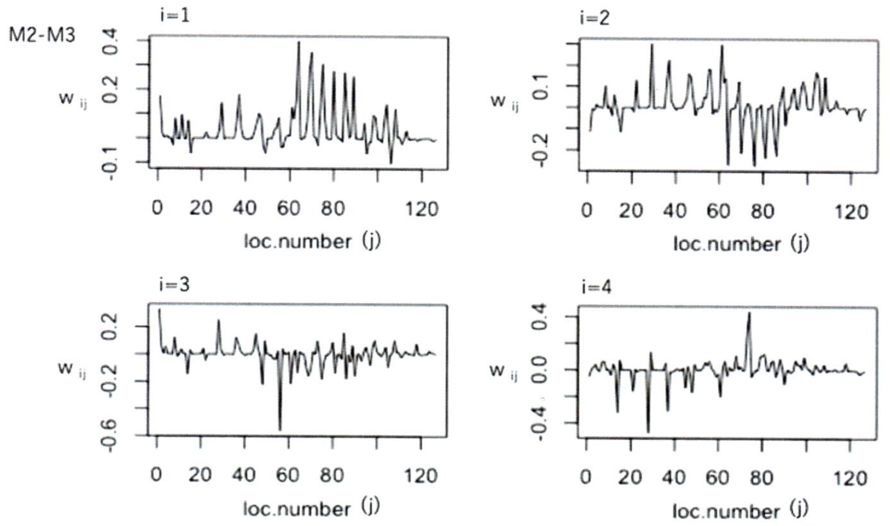

Fig. 6.22 Linear patterns of the correlation intensity (w_{ij}) of the locality cells shown in Fig. 6.4 of the mantle in the Japanese islands region

6.6 Correlated Seismicity Rates on the z_1–z_2–z_3 Diagram

Next, the diagrams of the z_1, z_2, and z_3 of the correlated seismicity rates for M1–2 of the crust will be investigated to identify the nature of clusters in these diagrams, as shown in Fig. 6.23. The mappings of them are very important in considering the dynamics of seismicity because huge amounts of data plots give us the sharp image of clusters which are identified to be segments of the time series noted later. The

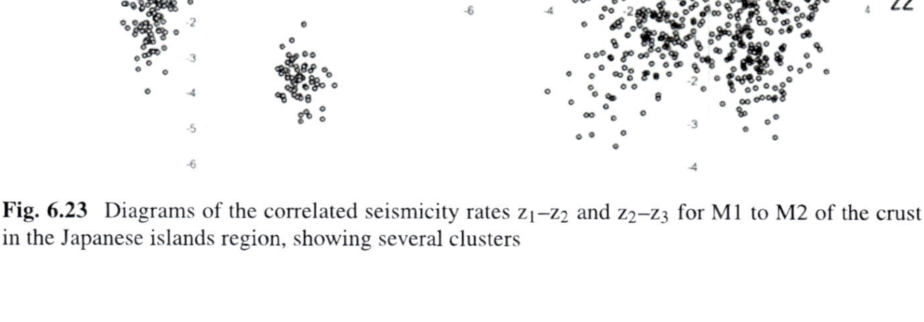

Fig. 6.23 Diagrams of the correlated seismicity rates z_1–z_2 and z_2–z_3 for M1 to M2 of the crust in the Japanese islands region, showing several clusters

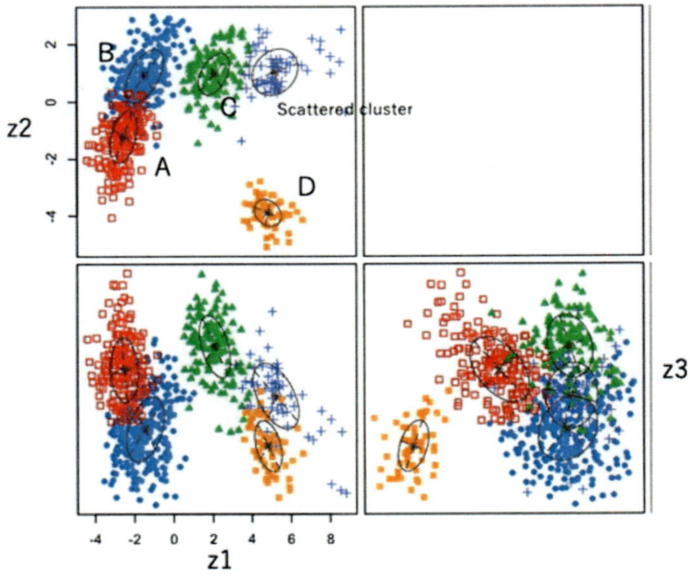

Fig. 6.24 Diagram showing the clustering of z_1, z_2, and z_3 data by means of "mclulst" package in R, showing the A, B, C, and D clusters and scattered cluster

probabilistic clustering of z_1, z_2, and z_3 was carried out by means of the application in *R* as "mclust" package, and 5 clusters can be identified in terms of multiple Gaussian distribution fitting as shown in Fig. 6.24. The model selection is based on the maximum of Bayesian information criterion (BIC; Bishop 2006) value. In this diagram, a scattered cluster at large z_1 and z_2 positions is clearly identified as the co-seismic and post-seismic signals and it is probably considered to be the overshoot by giant earthquake of Tohoku-Oki. Then, except for this scattered transient cluster, there should be considered to be four clusters in the z_1–z_2–z_3 diagram.

It is quite important that the clusters identified by probabilistic clustering method are identified also by the periods in the continuous-time sequence from 1998 to 2018: as is shown in Fig. 6.25, the cluster of the z_1 and z_2 in the period from 1998 to 2003 locates in the region of the third section, but it migrates toward the region of the second section from 2003 to 2008, and then it further migrates near the z_2 axis to 2011 before the Tohoku-Oki earthquake in the diagram. In the period from 2011 to 2015, the cluster with large variance by 2011 Tohoku-Oki eq. locates in the first section of the diagram and then it jumps to the fourth section in the period from 2015 to 2018. Considering that the correlated seismicity rates have multiple Gaussian noises that behave as random fluctuation in their time series, the clusters of every time intervals have dense portions to be called as the stable nodes of the correlated seismicity rates dynamics. On the other hand, the transient cluster has a wide scattering of z_1 and z_2 with larger variance than above stable nodes. Then it should be considered not to be the stable node but the transient node. These characterizations

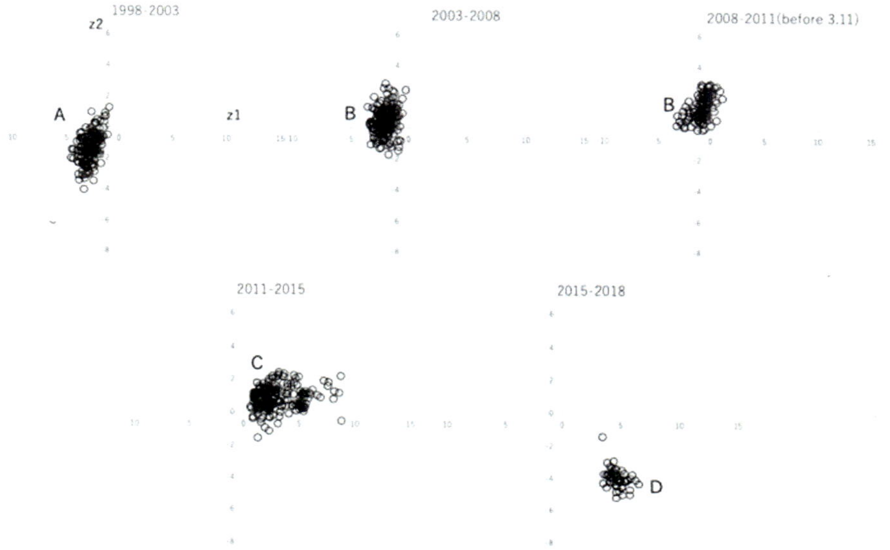

Fig. 6.25 Successive change of the correlated seismicity rates in the z_1–z_2 diagram, showing the transitions from A to B, B to C, and C to D clusters

of the correlated seismicity rate mapping in the migrating-averaged z_1–z_2 diagram as shown in Fig. 6.26 will be discussed again in the chapter of dynamics.

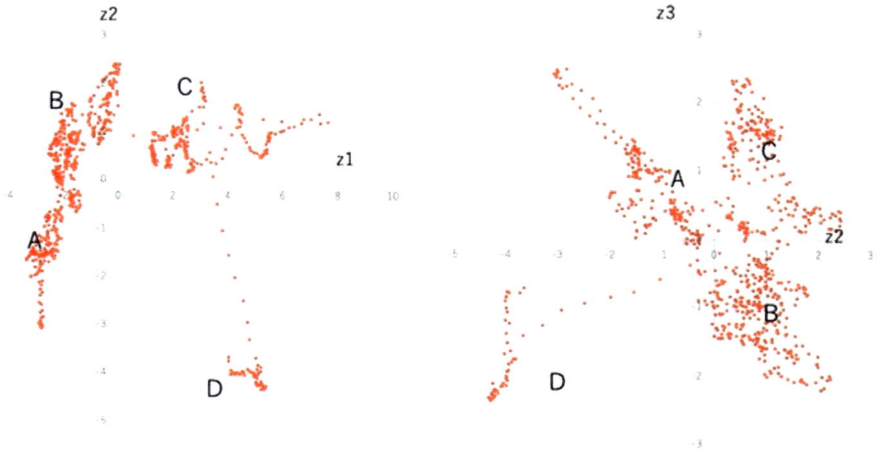

Fig. 6.26 Diagrams of the average correlated seismicity rates z_1–z_2 and z_2–z_3 of the crust in the Japanese islands region, showing several clusters, showing the A, B, C, and D clusters and scattered cluster

In the case of the mantle under the islands arc of Japan, the mappings are also complicated in the z_1 and z_2 diagram although there are five clusters by means of the probabilistic clustering method above as shown in Fig. 6.27. By above-mentioned stochastic clustering of z_1, z_2, and z_3 of the mantle, it suggests that the trend of the time series on this diagram displays the transition from one cluster to the other cluster as shown in Fig. 6.28. Although there classified into five clusters, the green cluster is not identical to the red cluster in Fig. 6.27, and therefore, four clusters should be considered as the available classification. It seems, then, important that the transitions from blue to purple cluster and from purple to red and orange clusters in Fig. 6.27 occur nearly discretely with increasing the time but these transitions are not associated with the giant earthquakes such as 2011 Tohoku-Oki as shown in Fig. 6.23.

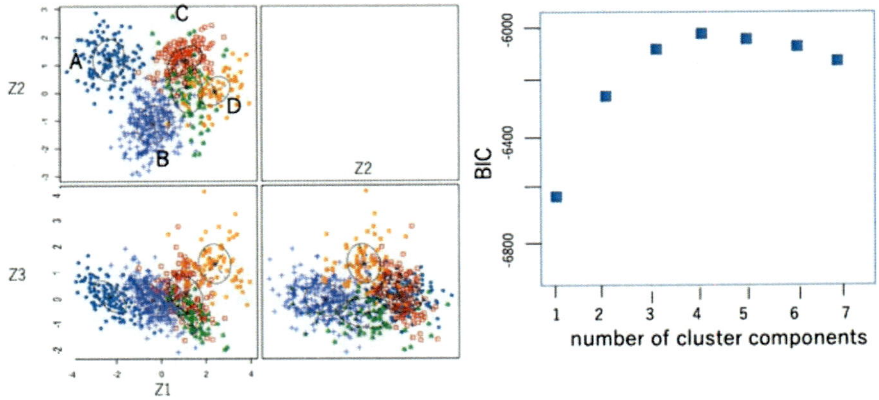

Fig. 6.27 Diagrams of the correlated seismicity rates z_1–z_2 and z_2–z_3 of the mantle (M1–M2) in the Japanese islands region, showing clusters of A, B, C, and D by means of "mclust" package in R

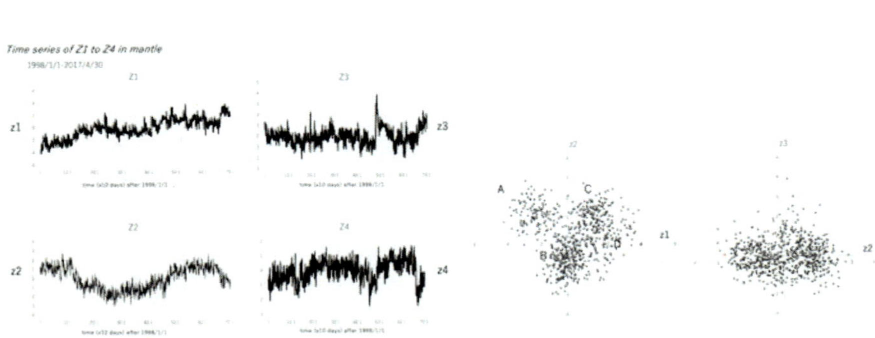

Fig. 6.28 Time series of the correlated seismicity rates z_1 to z_4 for M1–M2 of the mantle in the Japanese islands region from 1998/1/1 to 2017/4/30. Right; Diagrams of the correlated seismicity rates z_1–z_2 and z_2–z_3 (for M1–M2) of the mantle in the Japanese islands region, showing A, B, C, and D

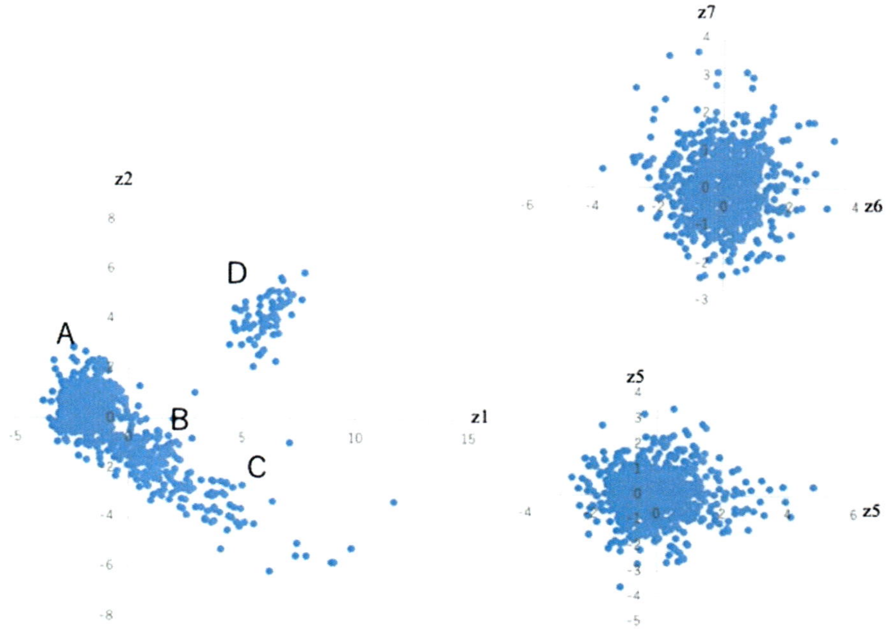

Fig. 6.29 Diagrams of the correlated seismicity rates z_1–z_2 for M2 to M3 of the crust in the Japanese islands region, showing clusters of A, B, C, and D. The right figures are the relations of z_5–z_6 and z_6–z_7 showing the random distribution patterns

The mapping of the correlated seismicity rates of M2 to M3 is displayed in the diagram of z_1–z_2–z_3 as shown in Fig. 6.29. It seems that there are three clusters (A, B, and C) but two of them are mutually overlapped though there are two dense cores of the clusters. It is probable that the clusters of the z_1 and z_2 in the M1 to M2 correlated seismicity diagram (A, B, and C in Fig. 6.26) correspond to the present overlapped cluster in above figure.

6.7 Coherency of Correlated Seismicity Rates Between Mantle and Crust

As stated in the previous sections, the correlated seismicity rates z_i of the crust are likely to those of the mantle in the Japanese islands region. Accordingly, the coherency of them should be investigated by means of the mappings of the z_1 or z_2 of the mantle and the crust as shown in Figs. 6.30 and 6.31. From these figures, it concludes probably that there are three clusters in the z_1 diagram of the mantle against crust: These clusters show the coherent trends in which the z_1 of the crust increases with increasing that of the mantle. On the other hand, three clusters are basically

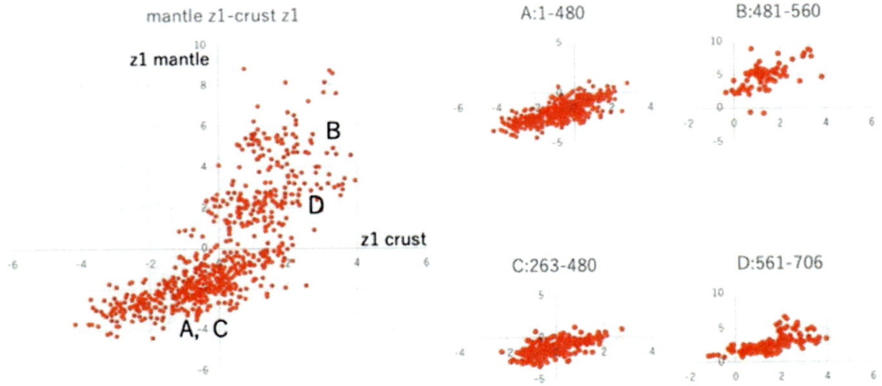

Fig. 6.30 Relation of the correlated seismicity rates z_1 for M1 to M2 between the mantle and crust in the Japanese islands region, showing the successive change from A to D with time in right figure

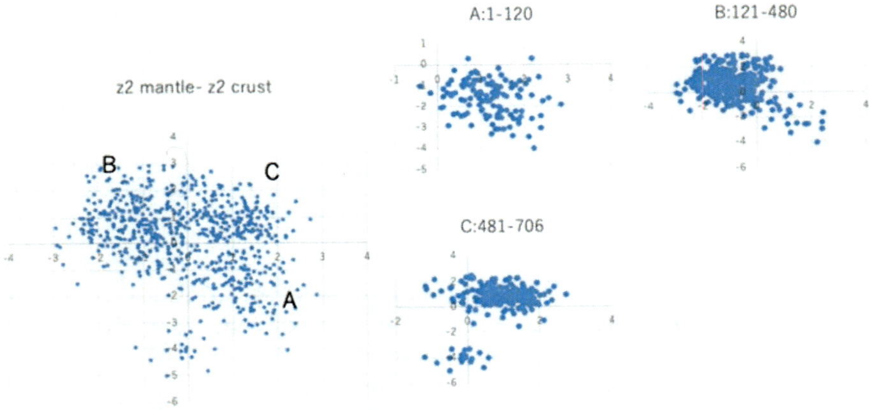

Fig. 6.31 Relation of the correlated seismicity rates z_2 for M1 to M2 between the mantle and crust in the Japanese islands region, showing the successive change from A to C with time in right figure

divided by the z_1 of the crust, so that the z_1 of the crust in three clusters increases discretely in the large z_1 regions of the mantle. In the same diagrams showing the plots of the different periods, the z_1 of both mantle and crust belonging to the time intervals from 1998 to 2011 and 2013 to 2016 (A and C in Fig. 6.30) make the cluster of low z_1 regions of the crust and within this cluster the plots in the early periods occupy the lower z_1 regions of the crust and mantle. The cluster B occupies the period between 2011 and 2013 but the cluster D does the period from 2016 to 2018. Furthermore, the z_1 values of the mantle and crust increase according to the time advancement in the same figures.

The second cluster over the low z_1 trend mentioned above in Fig. 6.30 is constituted with z_1 data from 2014 to 2018, though the third cluster over this is made up of the z_1

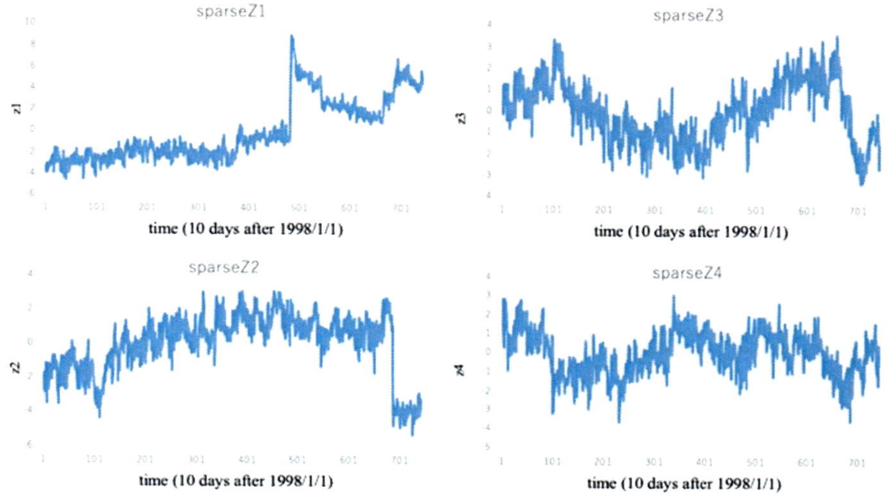

Fig. 6.32 Time series of the correlated seismicity rates z_1 to z_4 for M1–M2 by sparse PCA with 8 dimensions of the crust in the Japanese islands region from 1998/1/1 to 2018/8/28. This figure is the same as Fig. 6.12

data from 2011 to 2014, being just for the post-seismicity of Tohoku-Oki earthquake. Therefore, it is evident that there is a different situation of the correlated seismicity rates between the periods of 1998–2011 and 2011–2018.

On the other hand, the z_2 values of the mantle and crust are also plotted in the some different diagrams as the z_1 plotting as shown in Fig. 6.31. It is clear that there are three clusters in this figure, but the z_2 plots are scattered in the rather wide range in the diagram; but as is seen in Fig. 6.31, the rough trends from the region of A to C clusters through B cluster in the z_2 diagram. In addition, it is possible to consider the small cluster appears in the lower domain under the A cluster in the period from 2016 to 2017.

The seismicity rates matrix can be also investigated by the method of sparse modeling with LASSO using the "sparsepca" of R-packages. The time series and their attractor diagrams of the results are shown in Figs. 6.32 and 6.33, suggesting that all of them are very similar to those obtained by the deterministic PCA method described earlier. However, it seems that the annual variation seen in the z_2 time series can be more clear than that of the deterministic PCA modeling as shown in Fig. 6.32. Therefore, the annual variation will be discussed in the next sections.

6.8 Annual Variation of the Correlated Seismicity Rate

The subannual to biannual variations in the seismicity are recently found in non-volcanic tremors associated with slow slip events from the subduction boundary zones

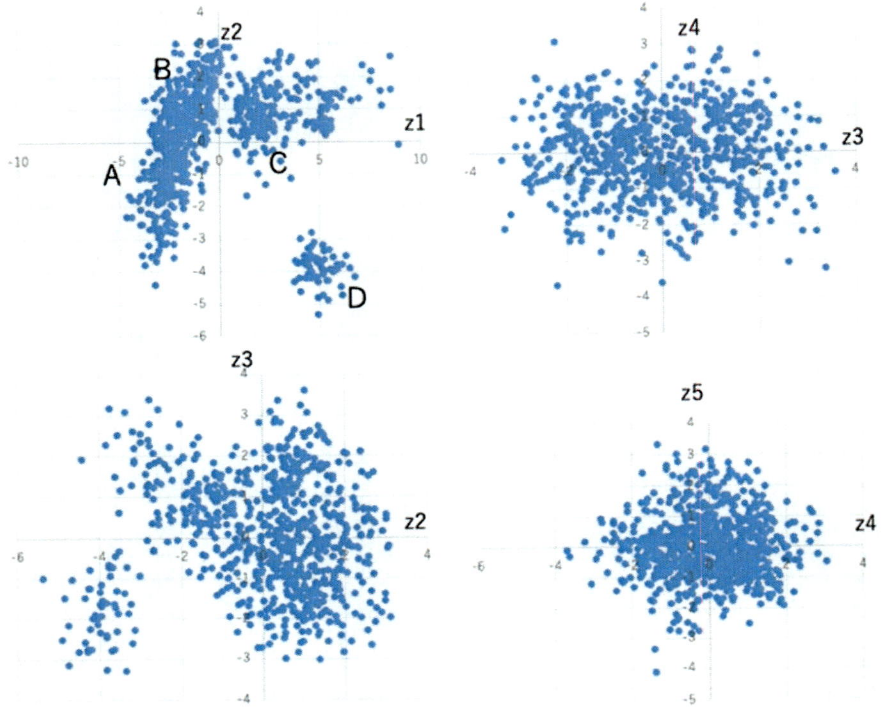

Fig. 6.33 Diagrams of the correlated seismicity rates z_1–z_2 and z_2–z_3 for M1 to M2 by sparse PCA with 8 dimensions of the crust in the Japanese islands region, showing clusters of A, B, C, and D

of Cascadia and SW Japan. In this study, it is sure that the correlated seismicity rates are apparently associated with the subannual to bi- to tri-annual cyclic fluctuations in the short-term fluctuations obtained from subtraction of general secular trends (Fig. 6.34) of the correlated seismicity rate time series as shown in Fig. 6.35. The short-term time series of the correlated seismicity rates in Fig. 6.35 suggests that there are cyclic fluctuation patterns in the time series subtracted by the 20 days migration average of the z_1, z_2, and z_3 time series. The near-annual periodic patterns are very distinguishable in the z_2 short-term variation rather than those of the z_1 and z_3 variations. In the z_2 short-term variation, the annual periodicity appears in the whole range of short-term time series from 1998 to 2018, though it seems to disappear in the range from 2003 to 2014 as shown in Fig. 6.35. The fast Fourier spectra of the short-term z_1 of the mantle and z_2 of the crust are shown in Fig. 6.36, thereby suggesting the sharp peaks at near-annual frequency. The frequency diagram of the peak-to-peak timescales is also indicated in the diagram (Fig. 6.37), indicating the annual variation of the correlated seismicity rates ranging from 0.5 to 2 years.

In order to check the synchronicity as seen in Fig. 6.35 of annual fluctuations between the mantle and the crust, the Lissajous diagram showing the phase shift

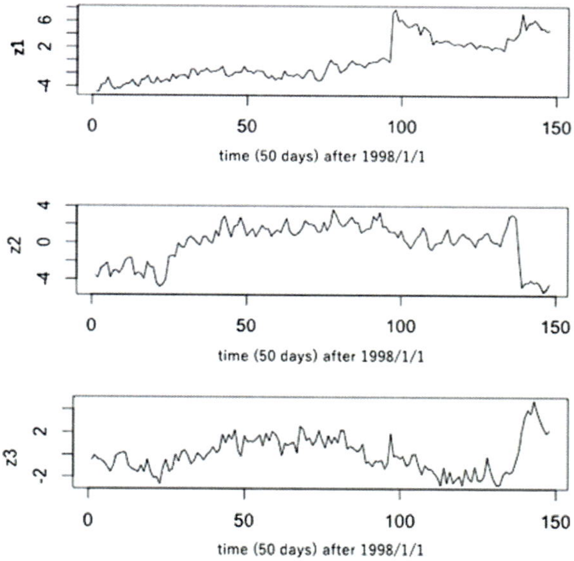

Fig. 6.34 Time series of the secular trends of the correlated seismicity rates in z_1 to z_3 of the crust in the Japanese islands region from 1998/1/1 to 2018/8/29

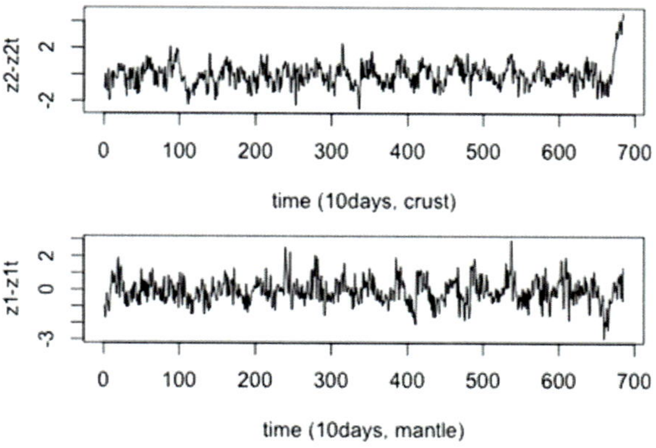

Fig. 6.35 Short-term variations of the correlated seismicity rates z_1 in mantle and z_2 in the crust subtracted by their secular trends (Fig. 6.34) showing the distinctive periodic variation with subannual to tri-annual patterns from 1998 to 2017

of the two short-term seismicity rates changes are presented in Fig. 6.38, but it seems that there is a little bit of phase shift between the mantle z_1 and crust z_2 short-term variations. It is, however, noted that two different trends in this diagram can be identified from the main dense trends, and these different trends also appear the synchronous change of the mantle and the crust with different intensity of the amplitude. The abnormal trend of the crust occurred at 2016–2017 but that of the mantle did at 2012–1013. However, it is not clear why these large amplitudes appear in these periods.

The short-term fluctuations of the M2–3 correlated seismicity rates z_i are also studied as shown in Fig. 6.39, thereby suggesting that the cyclic variation is relatively

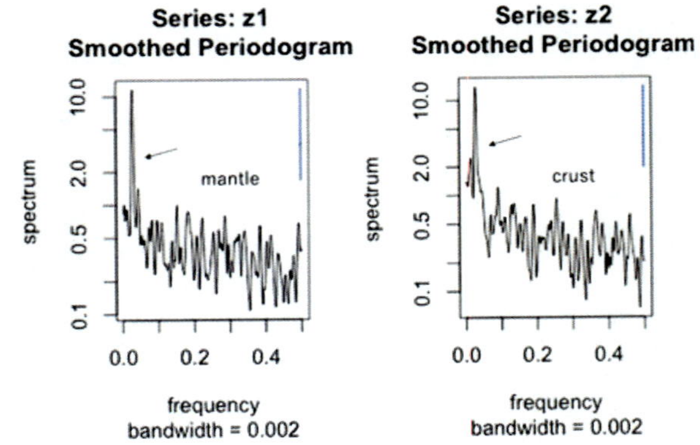

Fig. 6.36 First Fourier spectra of the short-term variations of the correlated seismicity rates z_1 of mantle and z_2 of crust in the Japanese islands region. Sharp peaks show the sub- to tri-annual periodicity

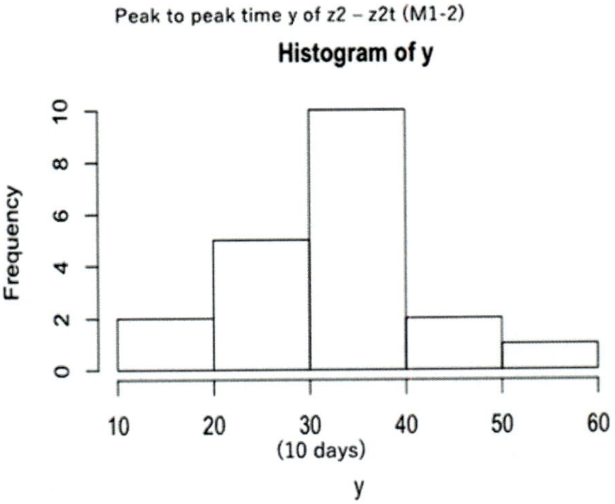

Fig. 6.37 Frequency diagram of the peak-to-peak time of the short-term variation of the correlated seismicity rate z_2 of the crust in the Japanese islands region

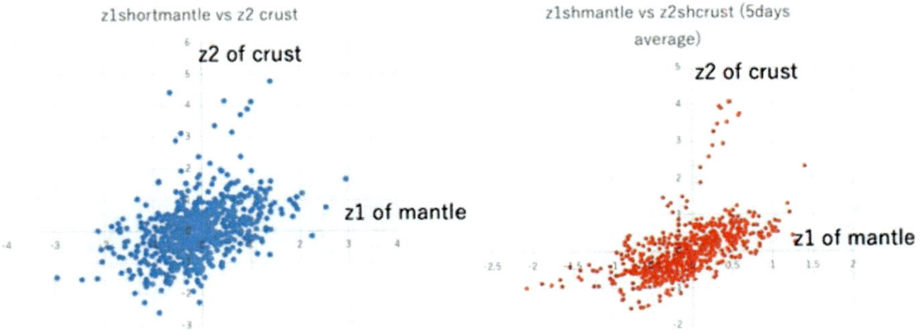

Fig. 6.38 Relations of the short-term variation in the correlated seismicity rates z_1 of the mantle and z_2 of the crust (left) and their 5 days average in the Japanese islands region (M1–M2)

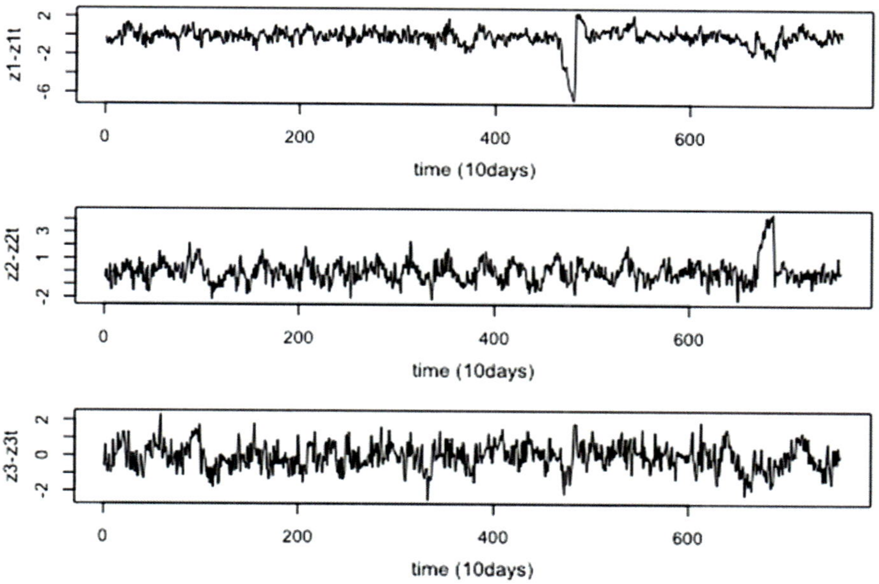

Fig. 6.39 Short-term variations of the correlated seismicity rates z_1 to z_3 for M2 to M3 in the crust subtracted by their secular trends, showing the distinctive periodic variation with subannual to tri-annual patterns from 1998 to 2017

clear in the time series of z_2 subtracted by its secular trend rather than those of z_1 and z_3 as being similar in the cases of M1–2 correlated seismicity rates (Fig. 6.40). In the time series of z_3, there seems to be a little sharp annual pattern in the times of 1998–2000 and 2015–2017. It may be noteworthy that there is noisy pattern in the z_3 time series before the time of the Tohoku-Oki 2011 earthquake.

The coherency of the z_2 time series between the M1–2 and M2–3 correlated seismicity rates is investigated in the Lissajous diagram described previously as shown in Fig. 6.41: The relationship between them is very obvious, and it indicates the distinctive coherency without phase shift. Therefore, it concludes that the subannual to tri-or biannual periodicity of the short-term fluctuations subtracted by general secular trends basically represent the characteristic correlated seismicity rate variations in the Japanese islands region, and it is probably due to the annual change in the population of seismogenic small to micro-shear cracks both in the crust and mantle. Besides, the coherent relations of population between these magnitudes of earthquakes are sure to be consistent with the Gutenberg and Richter law of statistic relation of earthquake populations described in the previous chapters.

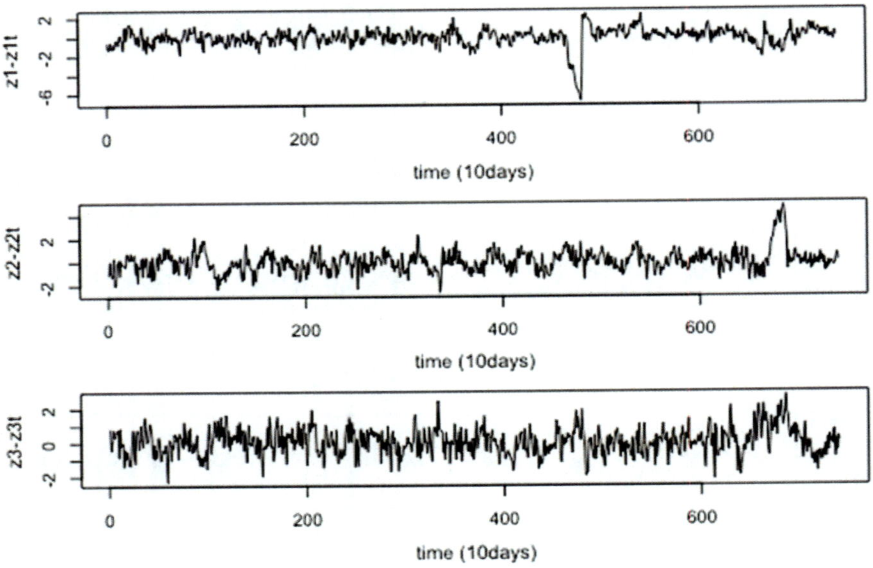

Fig. 6.40 Short-term variations of the correlated seismicity rates z_1 to z_3 for M1 to M2 in the crust subtracted by their secular trends, showing the distinctive periodic variation with subannual to tri-annual patterns from 1998 to 2017

6.9 Annual Variation of the Partial b-Value Time Series

The Gutenberg and Richter law of earthquakes population in the long term and large volume statistics has a characteristic parameter that means the slope in the diagram of the magnitude (logarithmic earthquake moment) and logarithmic frequency of earthquakes. This type of universal relationship between the moment magnitude and its frequency is responsible for the number density distribution of the shear crack size in the present system. If considering the stressed volume of the earth interior, there should be a critical Griffith crack whose size has the critical value between the growing and shrinking shear cracks. The critical size of the Griffith crack in rocks is determined by the elastic strain energy and surface energy ratio: For the former is larger than the latter, the shear cracks should grow, but for the case is reverse those should shrink. The critical size is considered to be about 1 m being corresponding to the micro-earthquake generating shear crack size of M1.

On the other hand, the shear cracks over the Griffith crack size, a, should propagate by the applied differential stress conditions in the relations of

$$v = \mathrm{d}a/\mathrm{d}t = K\sigma^n \tag{6.3}$$

in which σ and n are differential stress and its power (>50), respectively, and v is the velocity of the crack edge propagation (Anderson and Grew 1977). K is the stress

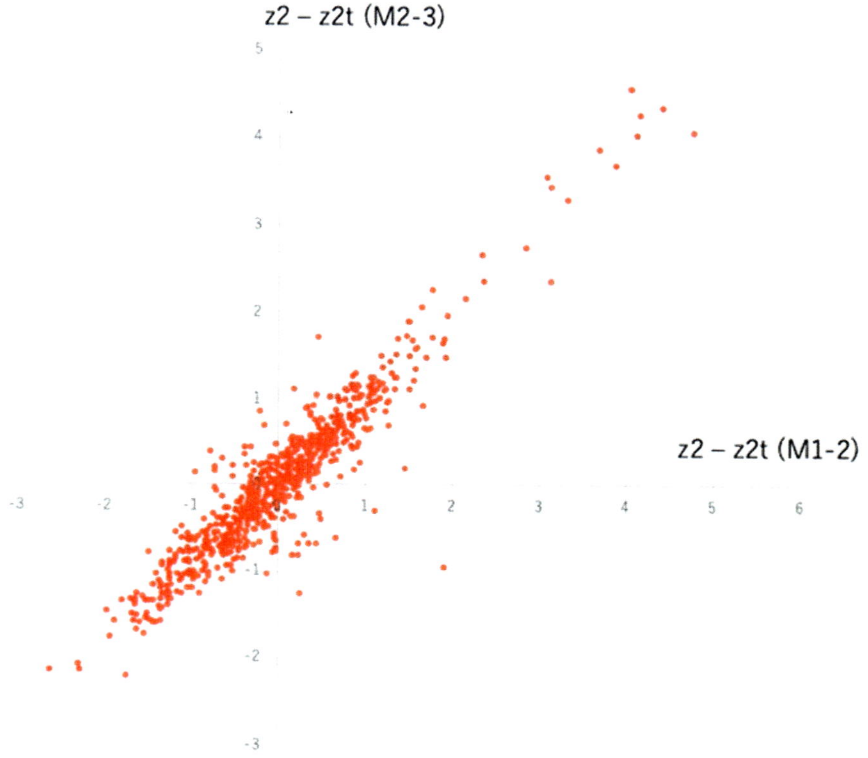

Fig. 6.41 Relation of the short-term variation of the correlated seismicity rates of z_2 for M2–M3 and for M1–M2 in the crust of the Japanese islands region

intensification factor proportional to root of the crack length a. On the other hand, the subcritical cracks also should grow slowly by the corrosion and hydration in the crack tips by H_2O diffusion and hydrated reaction of constituent minerals. As studied in detail using the experimental results by Ohnaka and Matsu'ura (2002), the slip velocity of the crack tip increases in the sigmoidal function in the logarithmic velocity and slip distance: At the initial state, the slip velocity increases very slowly, and after critical size of slip distance, it explosively increases to the second critical slip distance. Therefore, the first critical distance of slip is important for the generation of seismic waves and its size must be determined by the subcritical growth of crack, and it is proportional to the roughness indicator of the shear crack surface (ibid.).

Considering the water–rock interaction within the shear crack, the corrosion by dissolution of minerals at the crack tip should be associated with the sticking by mineral precipitation at the crack tip, and thus, the surface roughness indicator and the critical size of slip displacement mentioned above may be controlled by two terms of stress corrosion and mineral sticking rate as follows;

$$dL/dt = dl_c/dt - dl_s/dt \tag{6.4}$$

in which L is the length of crack, l_c is the crack tip enlargement length by stress corrosion, l_s is the crack tip shrinkage length by mineral precipitation.

If the stress corrosion is balanced with the mineral precipitation of above equation, the crack does not grow but because stress corrosion rate depends on the applied differential stress, the differential stress change should result in finite value of rate of the critical size L. Furthermore, these rates in the right hand of above equation depend upon the temperature and pressure in terms of activation energy and oversaturation. These mechanisms must change the growth rate of the critical size of displacement defined by Ohnaka and Matsu'ura (2002), and thus, there are variations in the critical size of slip in the crust and the mantle conditions.

According to Ohnaka and Matu'ura (2002), the critical slip size ranges from several micro-meters to several millimeters sizes and the critical length of nucleation zone of the high-speed slippage is also from several cm to several ten cm. It still is smaller than the shear crack size (1–10 m) of the micro-earthquakes of M1–2. On the other hand, the b-value of the Gutenberg and Richter law is the ratio of the earthquake frequencies of magnitude 1–2 against magnitude 2–3 or those of magnitude of 3–4 against 4–5, and so on. Therefore, it is suggested that the population of the micro- to small magnitude earthquakes is controlled by that of the critical length of the nucleation crack and also the critical slip size in the seismic regions except for the large earthquakes over magnitude of 5 corresponding the shear crack length about 1–10 km.

As discussed earlier, the critical length of slip size depends on the fluid pressure (water pressure), applied differential stress, pH of solution filling crack, mineral compositions of rocks, and temperature and pressure. Inasmuch as the pressure and temperature together with mineral compositions are considered to keep a nearly constant value during the seismic intervals, the time-variant characters are available for the fluid pressure, differential stress, and pH. Therefore, it is probable that the partial slopes of seismicity frequencies of magnitudes of 1–2 and 2–3 which are somewhat different with the global b-value of the GR law are indicative to the important state valuable for the seismicity change in the global and regional scales of the solid earth.

Here, the partial b-value is defined by the natural logarithmic ratio of earthquake frequencies of the magnitudes of 1–2 over those of 2–3 in this book as follows.

$$\begin{aligned} \text{b123} &= \log(1 + Nm_{12}) - \log(1 + Nm_{23}) \\ &= x_{ij,m_{12}} - x_{ij,m_{23}} \end{aligned} \tag{6.5}$$

in which Nm_{12} and Nm_{23} are number of seismic events ranged from M1 to M2 and M2 to M3, respectively, and $x_{ij,m_{12}}$ and $x_{ij,m_{23}}$ are their natural logarithmic values.

The time series of the partial b-values (b123) in every locality cells of the crust in Japanese islands region is displayed in Fig. 6.42 as like as the seismicity rates x_j. It is probable that the b-value shows the very noisy time series but that the general secular

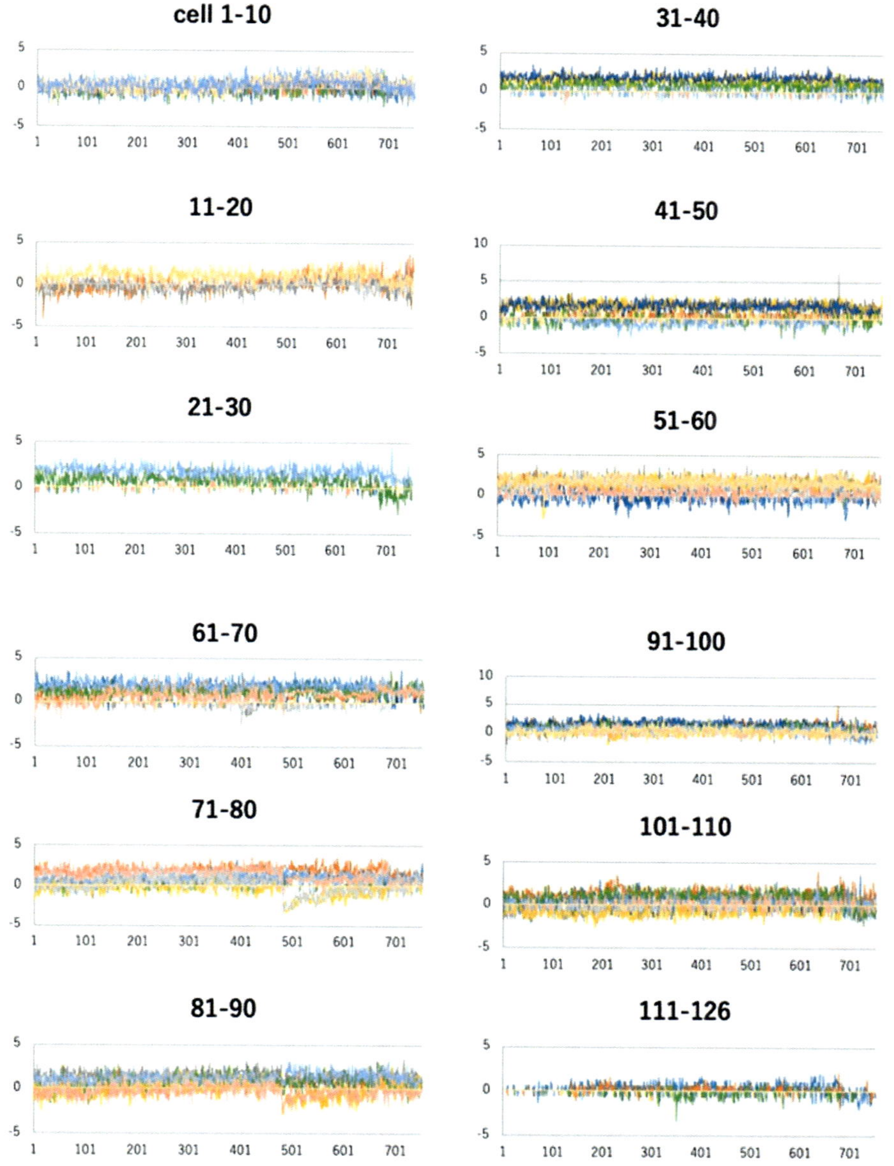

Fig. 6.42 Time series of the partial b-value (b123) variations of locality cells (1–126 in Fig. 6.4) in the crust of Japanese islands region from 1998 to 2018

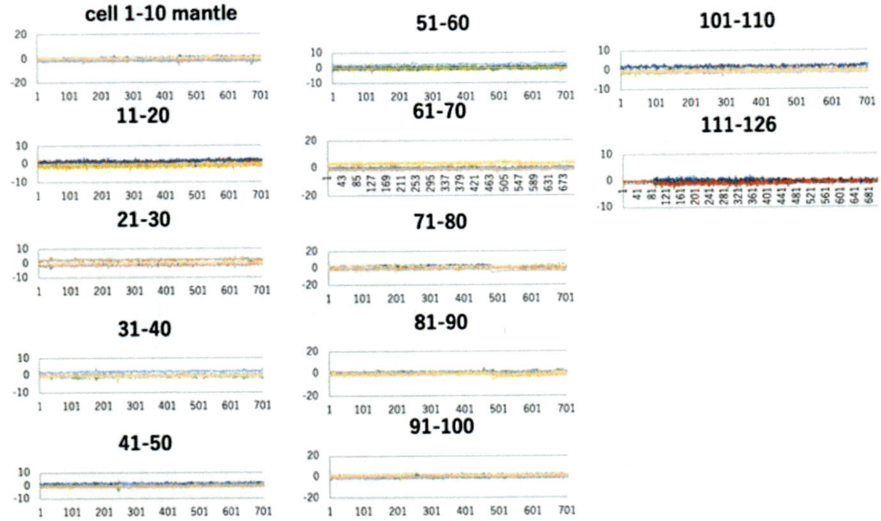

Fig. 6.43 Time series of the partial b-value (b123) variations of locality cells (1–126 in Fig. 6.4) in the mantle of Japanese islands region from 1998 to 2017

trends show somewhat large variations after 2011 even in the regions of NE and SW Japan. These secular trends of the partial b-values time series of many locality cells can be distinctively seen in the correlated partial b-value fluctuations in the following sections.

On the other hand, the partial b123 values time series of the mantle is shown in Fig. 6.43, suggesting that these time series of locality cells display the noisy general trends except for the cells of 122–126 of the Ogasawara islands indicating large variations of noisy time series. The seismicity near the Ogasawara islands seems to be unique because of rare seismicity around the present region. Considering the volcanic activity appearing in the west of Ogasawara islands, the seismicity and large variance of the partial b-value are derived from the magmatic activity in the upper mantle and the lower crust beneath the volcanic islands.

The correlated b-value variations of the crust and the mantle are shown in Figs. 6.44 and 6.45. The b123 time series of the crust are transformed to the correlated b-values of the locality cells, and their correlated b-value time series are shown in Fig. 6.44. It is very clear that the z_1 of the correlated b-value displays the gently wavy change from 1998 to 2011, but then suddenly increases at 2011 and then slowly decreases. It is noteworthy that there is a gradual increase of z_1 of the correlated b-value during three years before 2011 Tohoku-Oki giant earthquake, and in this period, there are cyclic fluctuations with nearly annual periodicity.

The z_2 variation of the correlated b-value b123 time series of the crust displays surprisingly the sharp cyclic fluctuation on its general trend as is seen in Fig. 6.44. The secular trend of this z_2 time series is constituted with four segments: The end of first segment is just on 2000 when Tottori intraplate earthquake occurred, and

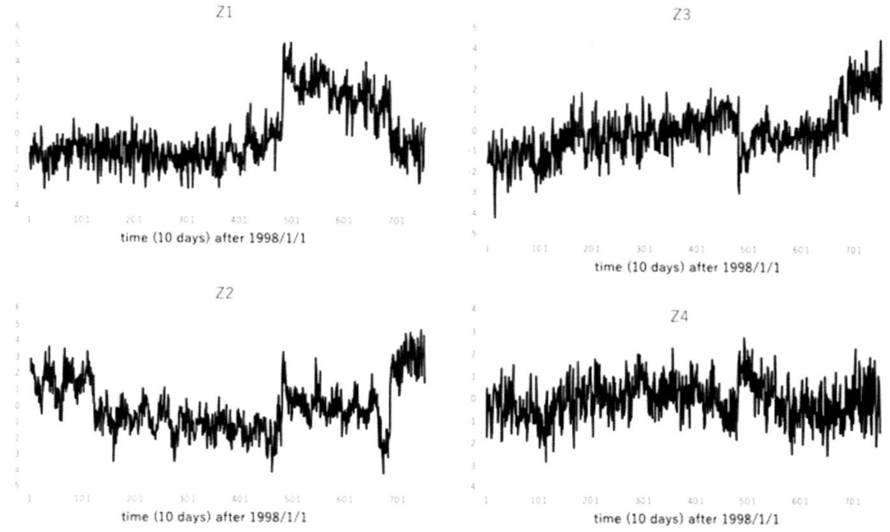

Fig. 6.44 Time series of the correlated partial b-value variations z_1 to z_4 for b123 (see text) of the crust in the Japanese islands region, showing the obvious periodic change in z_2 from 1998 to 2018

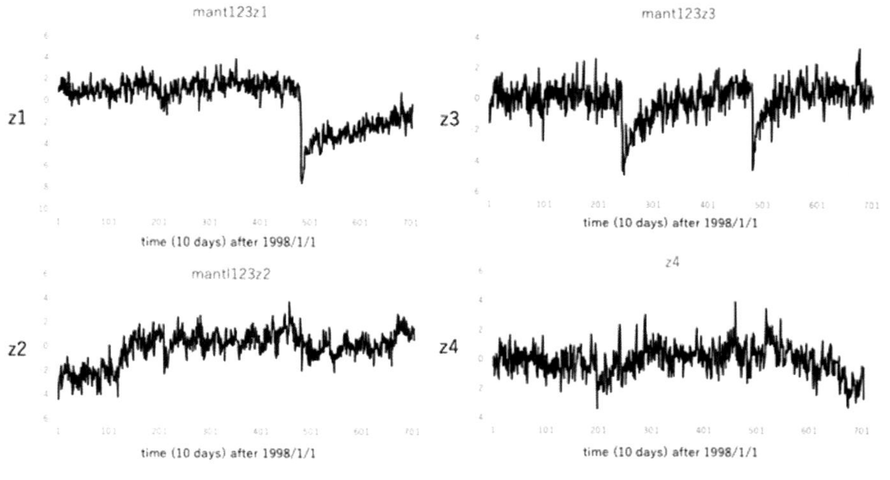

Fig. 6.45 Time series of the correlated partial b-value (b123) variations z_1 to z_4 for M123 (see text) of the mantle in the Japanese islands region, showing the obvious periodic change in z_2 from 1998 to 2017

that of the second one on 2011 of Tohoku-Oki giant plate boundary earthquake and that of the third is on 2016 of Kumamoto earthquake. It is, therefore, concluded that the segments are characterized by co-seismic correlated partial b-value change. As pointed out in the previous sections, the partial b-value, b123, may represent the mechanical states of the shear crack populations of the seismic regions, and thus, the correlated partial b-value change has an implication of regional environmental change on mechanical state of the seismic regions. Therefore, the z_2 time series of the crust shown in Fig. 6.44 must be important for inferring the large-scale mechanical change of the crust of inter-seismic or post-seismic periods. The several mechanical states of the seismic regions to be appeared are explained in the next sections.

On the other hand, the z_2 time series of the b123 of the mantle also appears the very similar pattern with that of the crust as shown in Fig. 6.45: at the time of Tottori earthquake z_2 of b_{123} in the mantle rapidly increased and at the time of the Tohoku-Oki earthquake it rapidly decreased with time as shown in Fig. 6.45. It also suggests that the z_2 of the b123 in the mantle increases and takes a peak before the Tohoku-Oki earthquake as same as the z_2 of the crust. Thus, it concludes that the b123 values of the crust changes coherently with that of the mantle in the Japanese islands region.

The correlation between the crust and mantle can be investigated by means of the diagram as shown in Fig. 6.46, suggesting the anti-phase lock between them with some degree of phase difference. The phase difference of the z_2 time series between the mantle and the crust seems to be less than 10–30 days. Their time series appears the trend different from the main one in the period from 2016 to 2017 in the above figure.

The annual fluctuations of the b123 time series are possibly clarified distinctively by the short-term variations by subtraction by the secular trends from the raw time series in the mantle and the crust as shown in Figs. 6.47 and 6.48. The secular trend of the b123 time series was calculated by mean of long-term migration average with

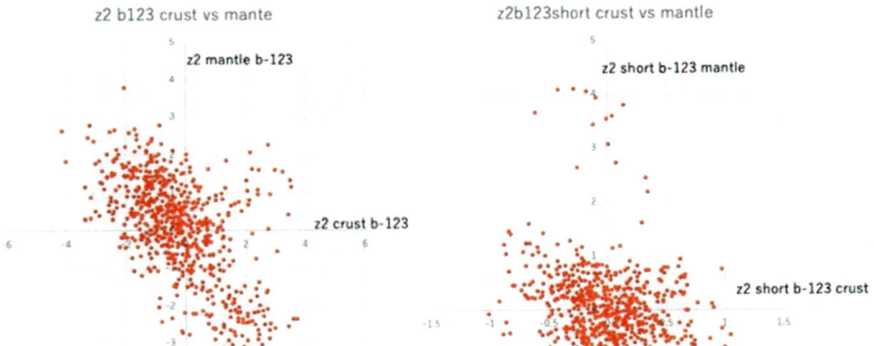

Fig. 6.46 Relations of the correlated partial b-values (b123) z_2 (left) and their short-term variations (right) for b123 between the crust and mantle in the Japanese islands region

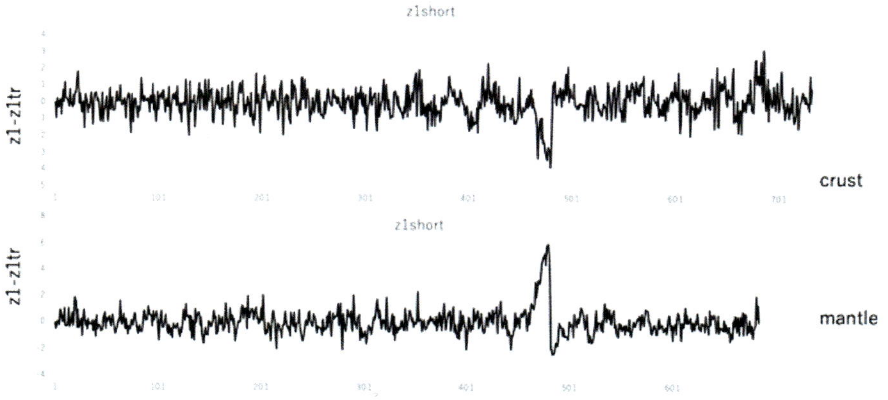

Fig. 6.47 Comparison of the short-term variations of the correlated partial b-value z_1 (b123) between the crust (1998–2018) and mantle (1998–2017) in the Japanese islands region

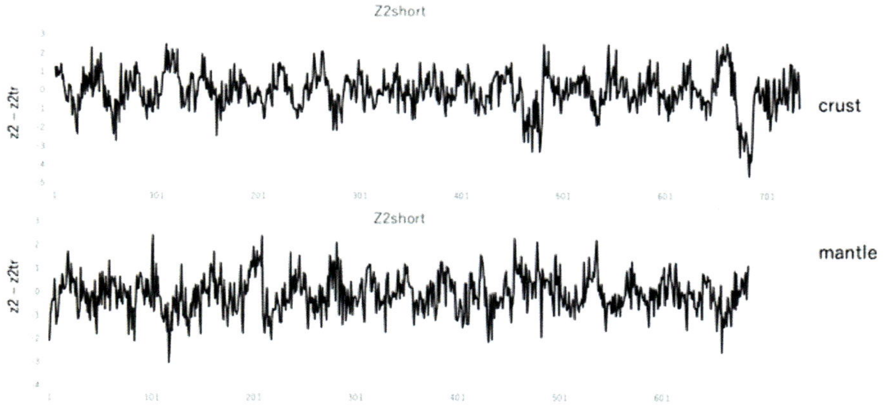

Fig. 6.48 Comparison of the short–term variations of the correlated partial b-value z_2 (b123) between the crust (1998–2018) and mantle (1998–2017) in the Japanese islands region

200 days. The figures indicate obviously that there are near-annual variations all through the studied period from 1998 to 2018. Particularly, the periodicity of the z_2 of the crust and z_1 of the mantle is very clear and their phase difference is less than about 10 days as seen in Figs. 6.47 and 6.48.

Therefore, the Fourier spectra of the z_1 and z_2 time series of b123 are studied by means of the FFT method using R-packages. The Fourier spectra of them are shown in Fig. 6.49, showing the distinctive peak of frequency about 0.05–0.07 which means the periodicity about 1–2 years. Furthermore, the time intervals of the peak to peak in the short-term fluctuation patterns of z_1 and z_2 time series are obtained about 0.5–2 years for both of the mantle and crust (Fig. 6.50), thereby concluding the annual periodicity of the b123 changes of both mantle and crust. It is also evident that the z_1

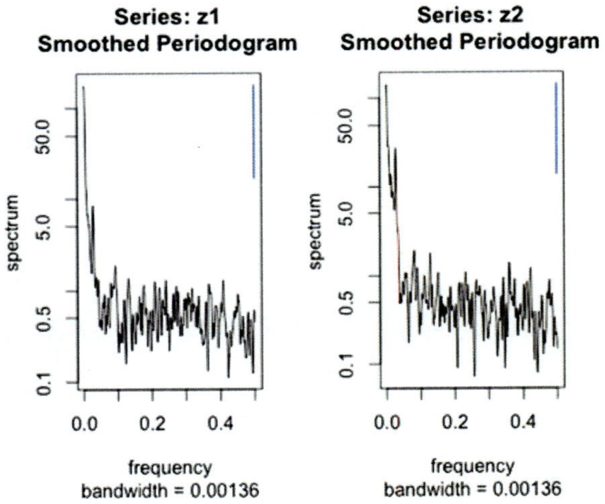

Fig. 6.49 First Fourier spectra of the short-term variations of the correlated b-value (b123) changes z_1 of mantle and z_2 of crust in the Japanese islands region. Sharp peaks show the sub- to tri-annual periodicity

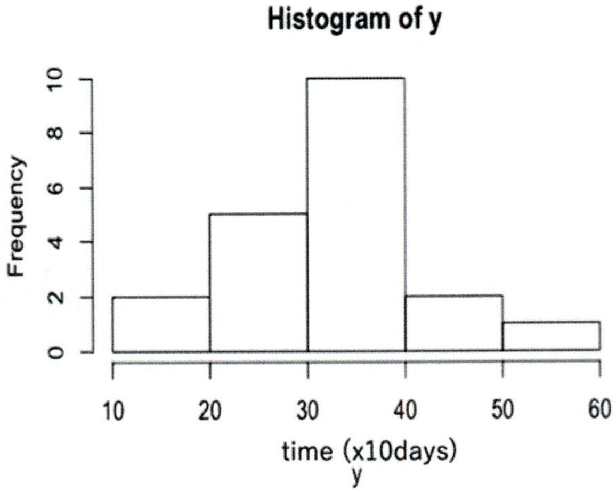

Fig. 6.50 Frequency diagram of the peak-to-peak time of the short-term variation of the correlated b-value (b123) changes z_2 of the crust in the Japanese islands region

of the mantle and the z_2 of the crust show that the short-term variations appear lose the obvious periodicity in the time segments between the clear periodicity segments.

The z_1 and z_2 values change during the studied period from 1998 to 2018 can be investigated by means of their incremental changes in the attractor diagrams. In Figs. 6.51 and 6.52, the data of every time steps of correlated b123 value in the crust and mantle from the studied period are plotted following the time ranges of each step, thereby indicating that the progressive change of the clusters of each one and each cluster looks like a Gaussian noise pattern. Thus, it is summarized the progressive change of b123 values in the whole Japanese islands region as indicated by Figs. 6.53 and 6.54 for the crust and mantle, and it is concluded that the z_1 and z_2 of the crustal b_{123} value go around anticlockwise three stable nodes in the studied periods involving

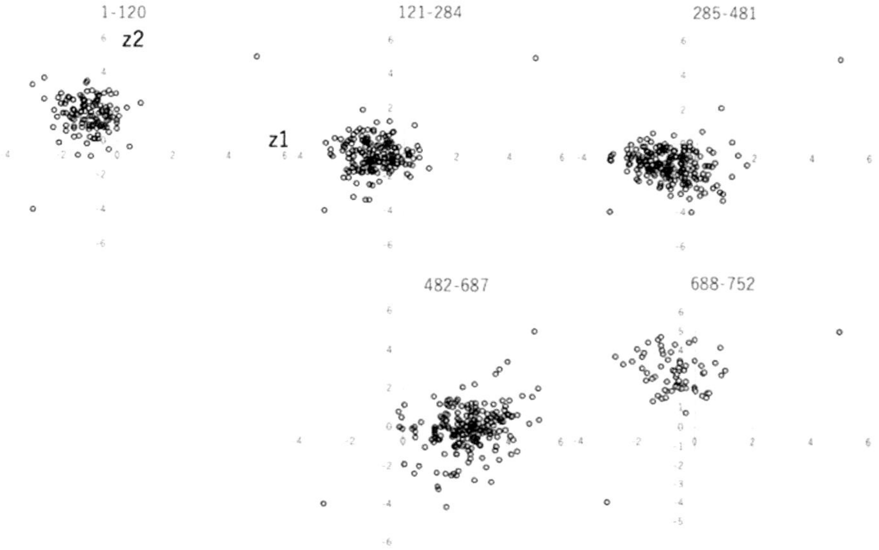

Fig. 6.51 Successive change of the correlated b-value changes z_1 and z_2 (for b123) of the crust in the Japanese islands region. Numbers are 10 days after 1998/1/1

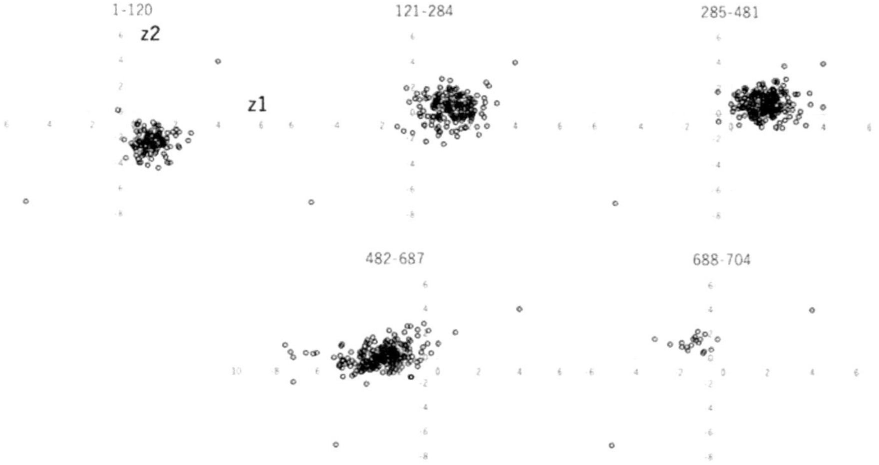

Fig. 6.52 Successive change of the correlated b-value changes z_1 and z_2 (for b123) of the mantle in the Japanese islands region. Numbers are 10 days after 1998/1/1

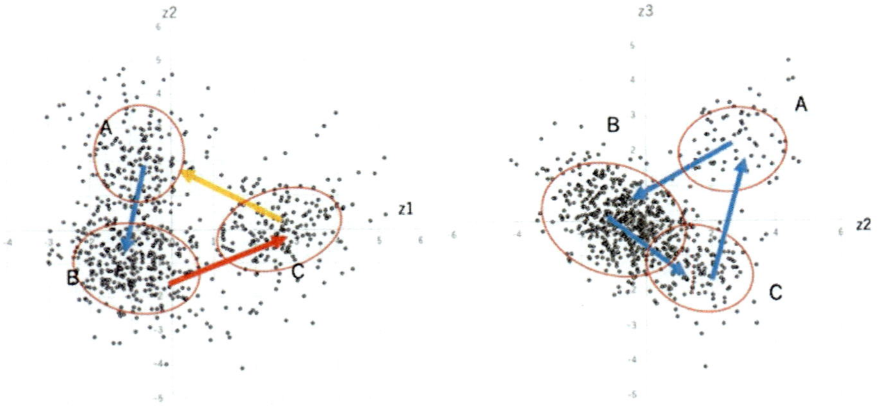

Fig. 6.53 Relations of the correlated b-value changes z_1 to z_3 (b123) of the crust in the Japanese islands region, showing the cluster transition from A to B, B to C, and C to A

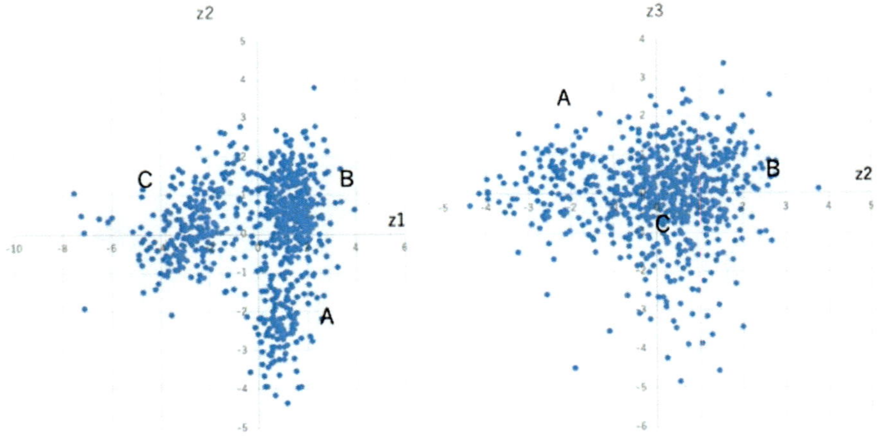

Fig. 6.54 Relations of the correlated b-value changes z_1 to z_3 (b123) of the mantle in the Japanese islands region, showing the cluster transition from A to B and B to C

the time of the Tohoku-Oki giant earthquake. On the other hand, the z_1 and z_2 move around also three clusters anticlockwise in the crust; but in the mantle (Fig. 6.54), the diagrams of z_1–z_2 and z_2–z_3 also show the anticlockwise rotation by three clusters.

The correlated local cell patterns governing the b123 time series of the z_1 to z_4 in the crust and mantle are shown in Figs. 6.55 and 6.56, respectively, revealing that the z_1 of b_{123} is constituted mainly with Okinawa and NE Japan districts though the z_2 with many locality cells from Okinawa islands to Hokkaido island regions. Considering that the z_1 of b123 time series shows the gentle cyclic variation and jump at Tohoku-Oki earthquake but the z_2 one does stepwise change and obvious

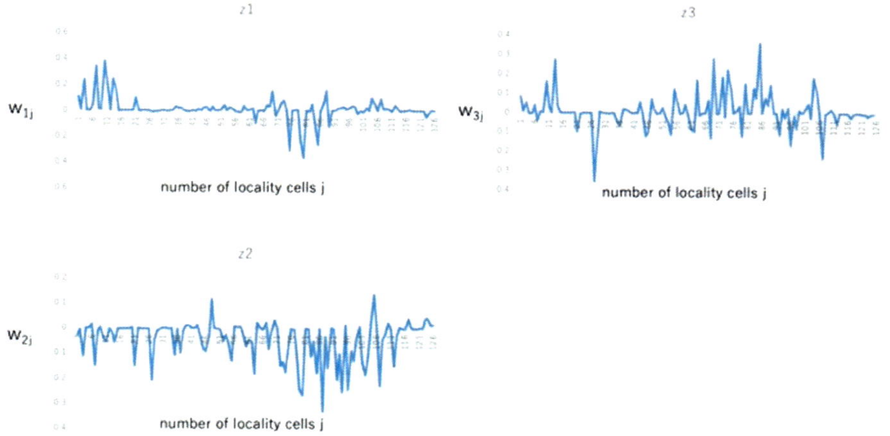

Fig. 6.55 Linear patterns of the correlation intensity (w_{ij}) for the correlated b-value (b123) changes z_1 to z_3 of the locality cells shown in Fig. 6.4 of the crust in the Japanese islands region

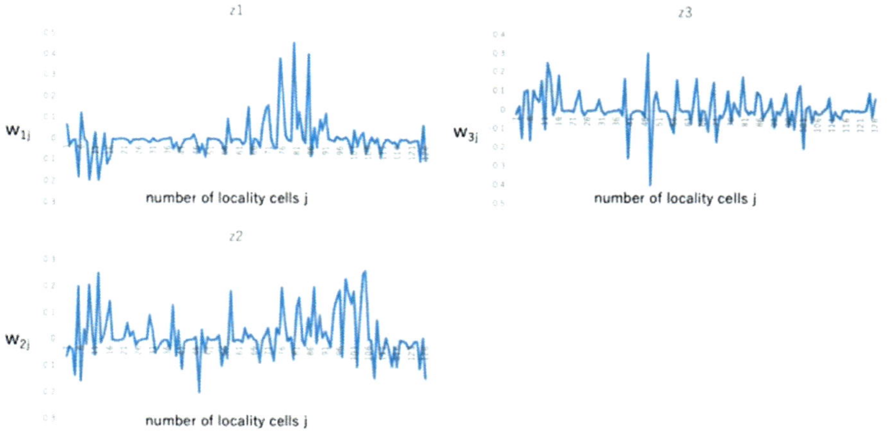

Fig. 6.56 Linear patterns of the correlation intensity (w_{ij}) for the correlated b-value (b123) changes z_1 to z_3 of the locality cells shown in Fig. 6.4 of the mantle in the Japanese islands region

annual fluctuation, the above-mentioned local cell patterns of the z_1 and z_2 mean that the areal correlation of Okinawa islands and NE Japan districts is intense and that of whole Japanese islands regions are characterized in the subannual to biannual variations of the b123 structure. It suggests that the mechanical states governing the b123 value of the plate boundary zones along the Japanese islands are synchronized in terms of the micro- to small seismic populations as shown in the diagram showing the relationship of b123, z_1, and z_2 between the mantle and crust (Fig. 6.57). However, the same tendencies of the b123 time series of the mantle and the crust can be identified in the whole districts of Japanese islands region as clarified previously, and thus, the

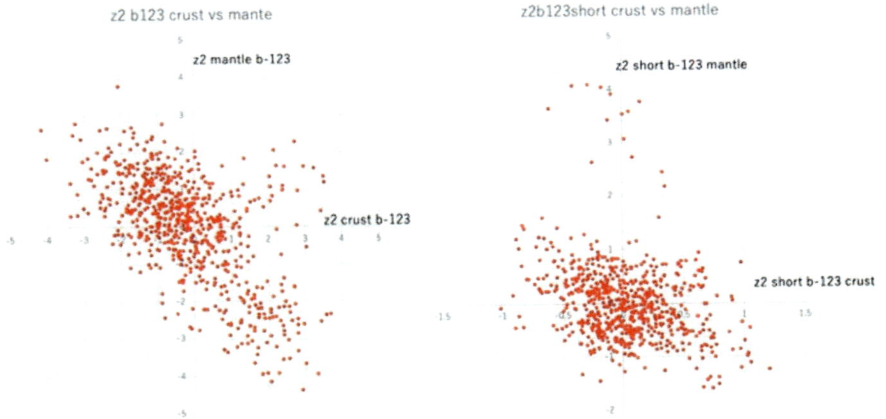

Fig. 6.57 Relations of the correlated partial b-values z_2 (left) and their short-term variations (right) for b123 between the crust and mantle in the Japanese islands region. This is same as those of Fig. 6.46

characteristic features of the z_1 and z_2 of the b123 time series should be responsible not for the surficial processes but for the slightly deep earth processes in the plate boundary zones such as plate subduction and solid tidal deformation (Nakata et al. 2008; Toriumi 2009; Tanaka 2012).

The partial b-value of b234 in these regions is also investigated by means of the transformation methods to the correlated z_1 to z_4 time series as shown in Fig. 6.58. It seems that the z_1 and z_2 of b234 display somewhat monotonous time series with large amplitude of noisy signals but that there is a sharp jump at the times of 2011 and 2016. The former jump is due to the giant Tohoku-Oki earthquake but the latter one is not certain to Kumamoto earthquake.

The time series of the z_2 appears also monotonous increase with noisy signal and rapid decrease on 2016 as like as the z_2, but the z_3 shows slight cyclic patterns with very noisy signals and sharp minimum on 2011 Tohoku-Oki earthquake. It is noteworthy that there are somewhat signals of annual variation likely to the z_2 of b123.

The short-term fluctuations of z_1 and z_2 of the b234 are possibly obtained by the subtraction of the z_1 and z_2 time series by secular trends of them, and these are shown in Fig. 6.59. It is probable that both short-term fluctuations are similarly associated with weak annual periodicity though these are very noisy.

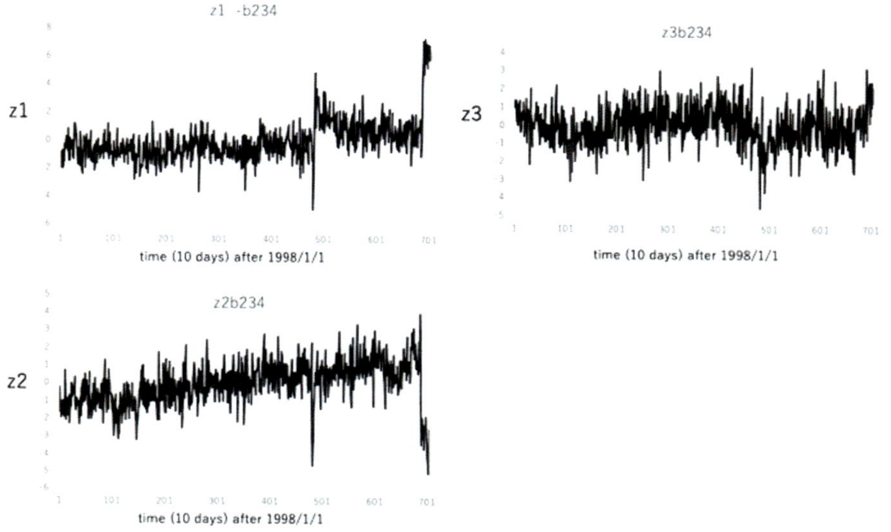

Fig. 6.58 Time series of the correlated partial b-value variations z_1 to z_3 for b234 (see text) of the crust in the Japanese islands region, showing the unclear periodic change in z_1 and z_2 from 1998 to 2017

6.10 Correlated Seismicity of Non-snowy and Snowy Regions of Japanese Islands

As mentioned in the previous sections, the z_1 and z_2 time series of the correlated seismicity rates of M1 to M3 is composed of secular variations with various long-term periodicities about 5–20 years and of short-term variations with subannual to biannual periodicities. These annual variations are observed in the geodesy time series by global network system of GPS and also in the gravity variations by satellite gravimetry method (Matsuo and Heki 2011). According to their results, it is usually suggested that the annual variations are often called as the seasonal variation and are due to the seasonal variations of loading and unloading of snow deposits or the seasonal variations of rainfall/precipitation derived from climatological processes. The former processes govern the periodic loading during winter and unloading during summer, but the latter should result in the maximum precipitation from August to September. It is clarified that the GPS displacements of every localities of the world are synchronous with the precipitation variations (Heki 2007). On the other hand, seasonal variations of satellite gravity data of several regions suggest the annual variations of the variations of average density of the crust or the mantle due to the absorption of H_2O from precipitation.

On the other hand, the correlated seismicity rates and also the correlated partial b-values display the distinctive subannual to bi- or tri-annual variations in the short-term variations, thereby suggesting that the periodicity of the satellite gravity by Heki

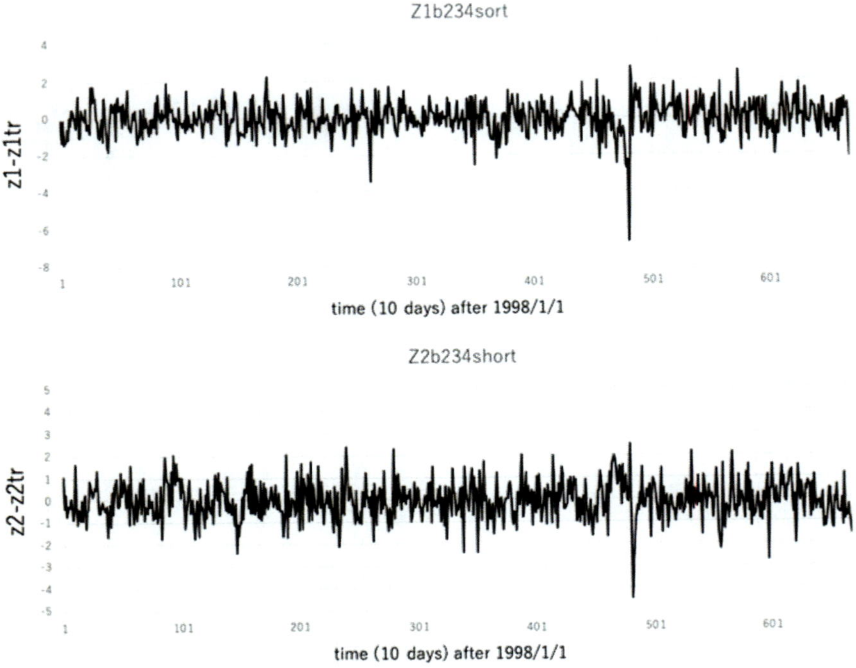

Fig. 6.59 Time series of the short-term variation of the correlated b-value changes z_1 and z_2 for b234 of the crust in the Japanese islands region showing the periodic change of subannual to tri-annual cycles

(2010) using GRACE data and network GPS data (Heki 2007) are not consistent with those of the correlated seismicity data studied here. Furthermore, the obtained periodicity of the correlated seismicity rates displays the coherency of the short-term variations including near-annual variations both in the mantle and the crust, showing that such periodicity should be derived from dynamic process coeval in the deep crust and mantle.

In order to check the possibility of the snowy loading and unloading resulting the correlated seismicity rates variations likely to the satellite gravity and network GPS annual variations, the comparison of the annual variations of them of the snowy and non-snowy regions in Japan is investigated by downscaling method for the data sets of whole Japanese islands region: The downscaling method is that whole data set is divided into two subsets and the one set is constituted with the locality cells of snowy regions and the other with those of non-snowy regions as shown in Figs. 6.60 and 6.61.

In these figures, it shows that the variations of the correlated seismicity rates in the non-snowy resemble those in the snowy regions, thereby suggesting that the near-annual variations of the correlated seismicity rates are not governed by the snow loading and unloading mechanisms unlikely to the satellite gravity and network GPS

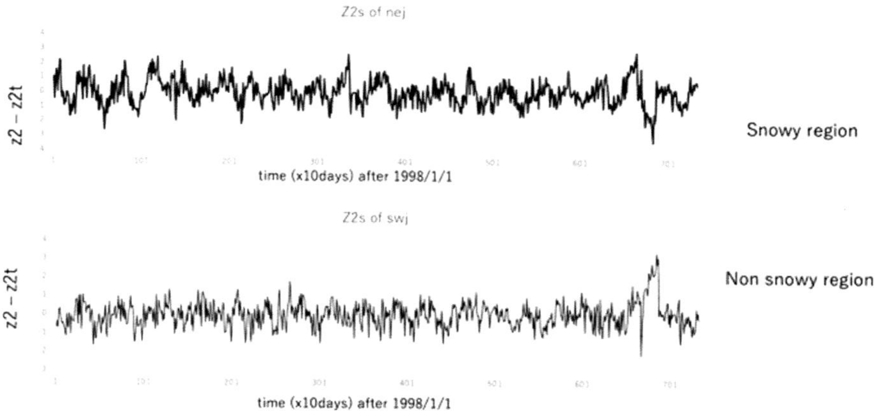

Fig. 6.60 Comparison of the short-term variations of the correlated b-value (b123) changes z_2 of the crust between the snowy region (NEJ) and non-snowy region (SWJ excluding Japan Seaside) in the Japanese islands, suggesting the same cyclic change

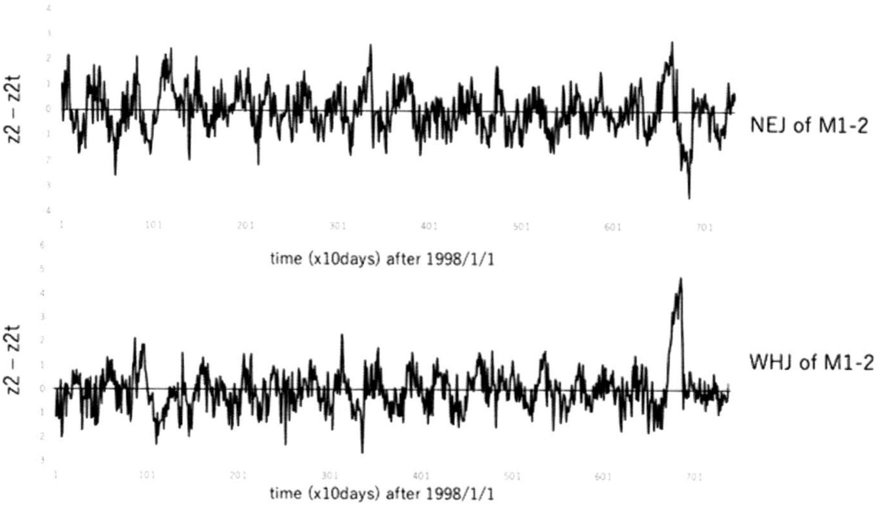

Fig. 6.61 Comparison of the short-term variations of the correlated b-value (b123) changes z_2 of the crust between the whole Japan region and Northeast Japan region, suggesting the same but anti-synchronous change

annual variations. On the other hand, it remains that the precipitation loading is possibly the reason of the near-annual periodicity of the correlated seismicity rates. However, the peak of the rainy precipitation of the Japanese islands regions is on September and its second peak on June, and it does not coincide with the peak of the correlated seismicity rates as mentioned previously. Thus, it may conclude that the

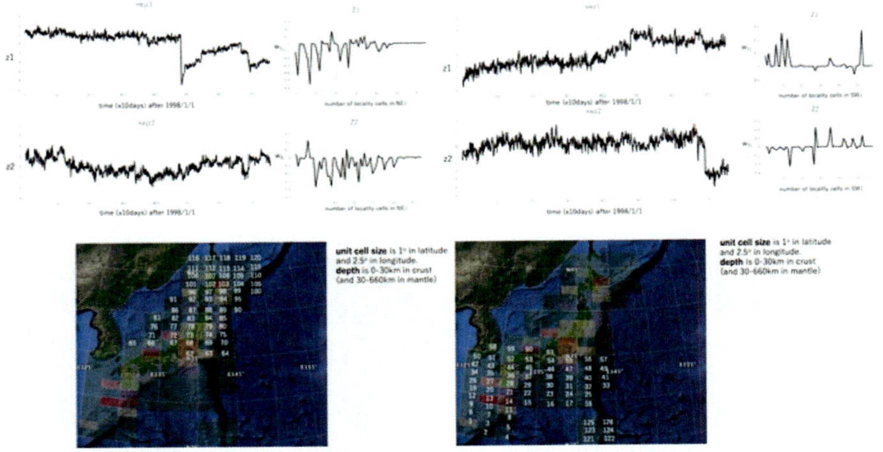

Fig. 6.62 Comparison of the correlated seismicity rates z_1 (top left) and z_2 (top right) for M1 to M2 and related correlation intensity of the locality cells between Northeast Japan (top) and Southwest Japan (bottom). Maps of the locality cells in the Northeast Japan (bottom left) and Southeast Japan (bottom right) are shown

cyclic variations of the studied correlated seismicity rates are not responsible for the snow and precipitation loading and unloading.

On the other hand, the z_1 and z_2 time series of the correlated seismicity in the snowy NE Japan and Japan Seaside, and non-snowy SW Japan are displayed in Fig. 6.62, showing that there is a difference of the z_2 patterns between them. It shows that the z_2 time series has rapid down at the 2016 in the non-snowy SW Japan, although there is no signal of such a rapid down in the z_2 time series of snowy NE Japan and Japan Seaside. It is probable that such difference in the z_2 patterns is derived from the rapid increase of seismic activity around the central Kyushu area.

The near-annual variations are known in the slow slip events appeared in the region along the subduction plate boundary as reviewed by Schwartz and Rokosky (2006). They summarized the repeated cycles of the SSE and the non-volcanic tremors. Particularly, the 1.5 years periodicity of the SSE associated with tremors avalanches is clearly observed in the region of Cascadia plate boundary. The SSE and related tremor activity are observed in the Nankai Trough plate boundary zone and their occurrence look like those of the Cascadia plate boundary. The periodicity of the SSE and related tremors, however, varies in the range from several hundred days to several years, being different from the annual periodicity of the satellite gravity and the network GPS variations. On the contrary, their periodicity larger than annual variation of the SSE seems to be consistent with the correlated seismicity rates variations studied in the book. So, it should be discussed later that the coherency between the SSE and correlated seismicity rates variations studied here are understood as the unified mechanical process along the plate boundary zone.

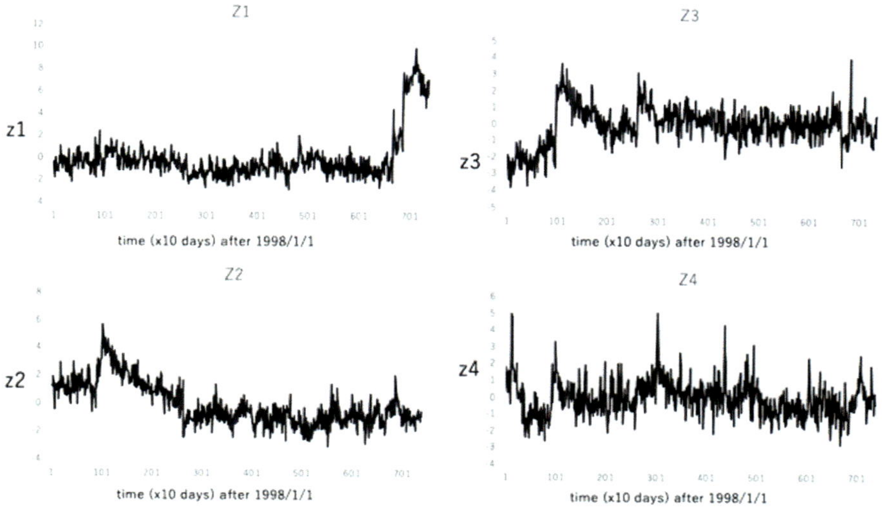

Fig. 6.63 Time series of the correlated seismicity rates z_1 to z_4 for M1–M2 of the crust of Southwest Japan excluding Okinawa area from 1998 to 2018

It may be noticed that the sudden change of the correlated seismicity rates z_2 on 2016 is lack in the NW Japan though it appears in the SW Japan, thereby suggesting that the sudden change of the z_2 time series above seems to be due to the change of the mechanical states in the SW Japan on 2016. However, the timing of this rapid change does not coincide with those of the large Kumamoto earthquake and other known earthquakes. It is, therefore, still of the unknown phenomenon.

Downscaling processing of the data matrix from the SW Japan is also investigated in order to clarify the behavior of the Philippine Sea plate along the plate boundary of Ryukyu trench and the Nankai Trough. In Fig. 6.63, the correlated seismicity rates of SW Japan excluding the locality cells of regions along the Sea of Japan are shown, and thus, the figure suggests that the time series of the z_1 displays a weak undulation from 1998 to 2016 and sudden increase after 2016. On the other hand, the z_2 time series shows abrupt increase on 2000 Tottori intraplate earthquake and it decreases gradually to 2017. The time series of the z_3 and z_4 appear almost always the noisy patterns after 2000.

On the other hand, the time series of z_1 for the data matrix of SW Japan excluding the data of Kyushu and Okinawa (Fig. 6.64) seems to show the rapid increase on 2000 and 2016 and slight change on 2004 at Kii-Oki earthquake, but that of the z_2 displays gradual increase from 1998 to 2016 and slight change at that time. There seem to the noisy fluctuations and jumps on 2000 and 2004 at large earthquakes in the z_3 and z_4 variations in the figures. Furthermore, the correlated seismicity rates of z_1 and z_2 time series are shown in Fig. 6.65 in the case of the data excluding the

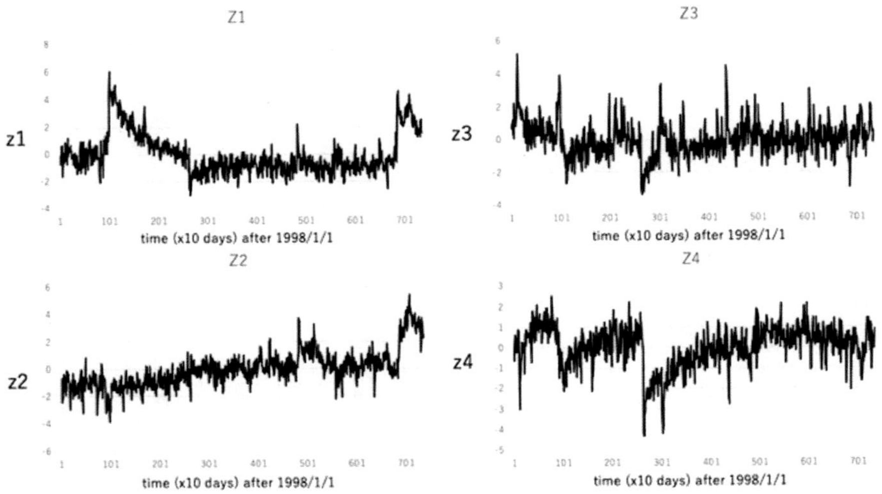

Fig. 6.64 Time series of the correlated seismicity rates z_1 to z_4 for M1 to M2 of the crust in the Pacific coast of SW Japan excluding Okinawa and Kyushu area from 1998 to 2018

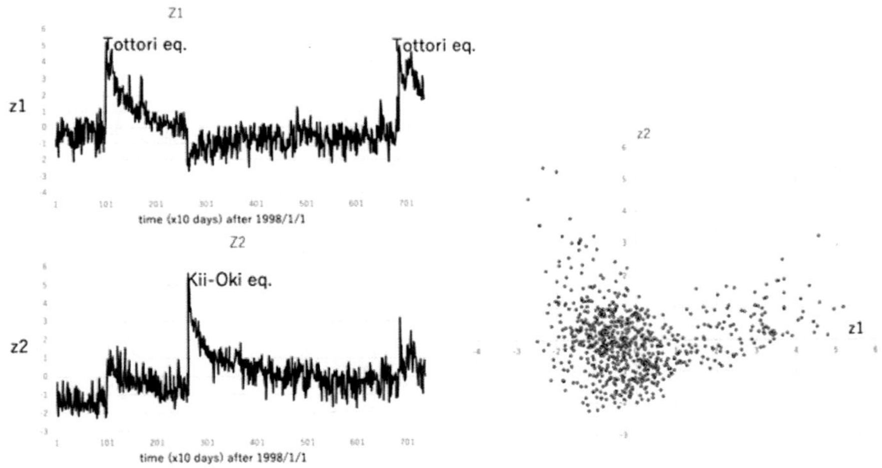

Fig. 6.65 Time series of the correlated seismicity rates z_1 to z_4 for M1–M2 of the crust in the Nankai Trough area of SW Japan excluding Kanto area from 1998 to 2018. The right figure displays the diagram of z_1 and z_2 of this area showing the single cluster with scattered spines

Kanto area, displaying the sharp peaks at 2000 Tottori earthquake, 2004 Kii-Oki, and 2016 Tottori earthquake.

In addition, the contribution of the locality cells in terms of the correlated seismicity is given in Fig. 6.66, and the figure implies that the weight coefficients of the z_1 component display the mode of contribution of many locality cells including Okinawa, Kyushu, Shikoku-Oki, and Kii-Oki. On the other hand, the weight coefficients diagram shows the contribution of Kanto, Okinawa, N Kyushu, and Tottori, suggesting the different modes of correlation of seismicity rates z_2 to z_4 from that of the z_1.

These modes of different correlation of seismicity rates of locality cells are possibly indicated by the diagram of z_1 and z_2 relations as shown in Fig. 6.67. It is suggested that the gradual change from the large z_2 and small z_1 cluster appears on 1998–2000 and then the correlated cluster changes toward the small z_1 and z_2 regions and then it jumps to the large z_1 and small z_2 region. The clusters of the z_1 and z_2 in each phase look like a Gaussian fluctuation as seen in above figures.

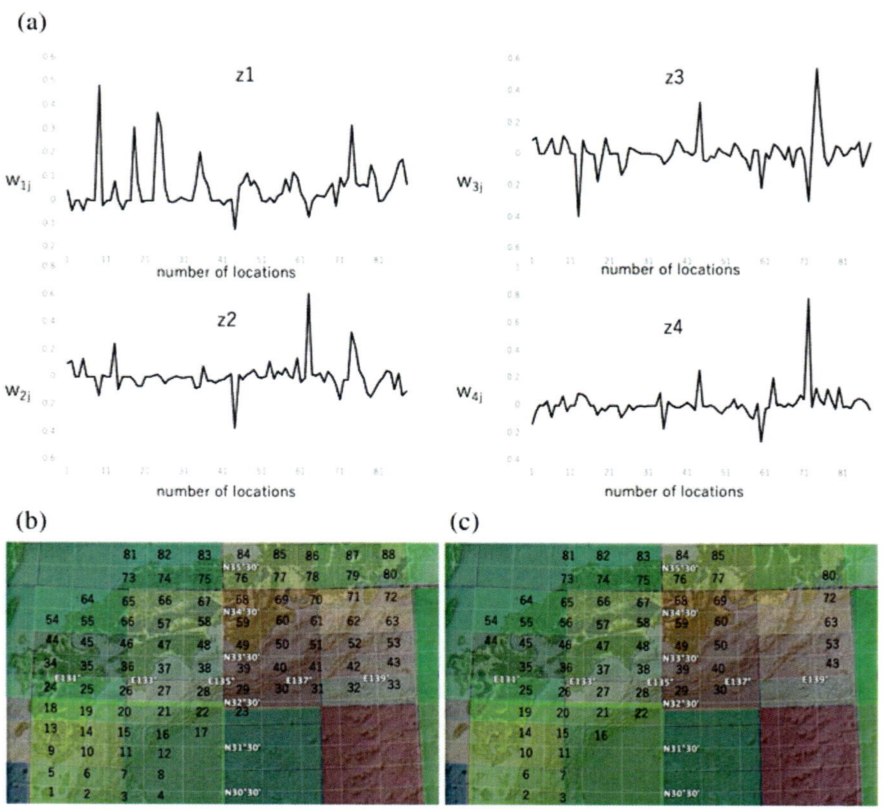

Fig. 6.66 **a** Linear patterns of the correlation intensity (w_{ij}) for the correlated seismicity rates z_1 to z_4 of the locality cells in the SW Japan. **b** Map showing the locality cells in the detailed data matrix of Southwest Japan (SWJ). **c** Map showing the locality cells in the data of the Nankai area excluding Kyushu and Kanto area of the Japanese islands region

Fig. 6.67 Diagrams of the correlated seismicity rates z_1–z_2 and z_2–z_3 for M1 to M2 of the crust in the SW Japan, showing clusters of A, B, C, and D

Stochastic analyses of these clusters are obtained as Fig. 6.68 by means of "mclust" in "RStudio" mentioned previously.

In the case of the downscaling to the Nankai Trough excluding Kyushu and Kanto regions, the correlated seismicity rates z_1 and z_2 change the patterns of time series from those of the SW Japan. Figure 6.65 indicates that there are clear peaks of the z_1 at the time of Tottori earthquakes on 2000 and 2017, and their peaks are followed by gradual decreases during 5–7 years. On the other hand, there are like patterns in the z_2 time series involving the peak at the 2004 Kii-Oki earthquake of which the post-seismic z_2 pattern shows duration of about 7–10 years as shown in Fig. 6.64. It is noteworthy that there seem to be a stable cluster and transient clusters in the z_1

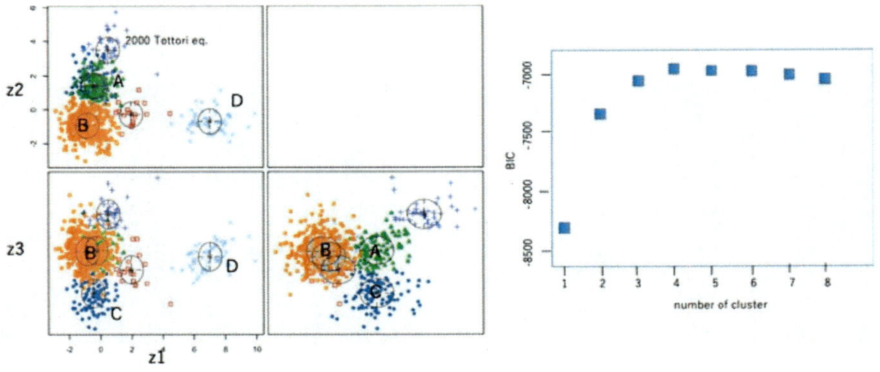

Fig. 6.68 Diagrams of the correlated seismicity rates z_1–z_2 and z_2–z_3 of the crust (M1–M2) in the SW Japan. Right figure shows the BIC diagram against the number of clusters, showing clusters of A, B, C, and D by means of "mclust" package in R

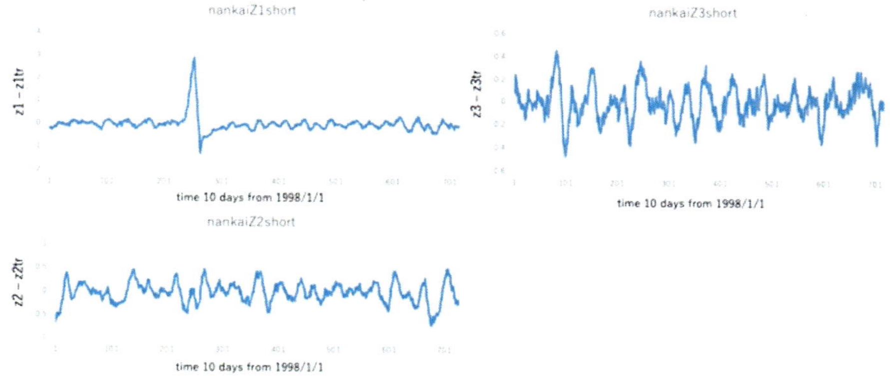

Fig. 6.69 Time series of the short-term variations of the correlated seismicity rates with random noise addition z_1 to z_3 in the Nankai region, showing the periodic change with 1–3 years cycles

and z_2 diagrams in Fig. 6.65. Furthermore, it seems that the transient clusters can be divided into two types: One displays the decreasing trend from large to small z_1, but the other does the decreasing one from large to small z_2. The latter is in the case of Kii-Oki plate boundary earthquake and the former in the cases of Tottori intraplate earthquakes, and thus, it may be possible that the one type transient trend after the Kii-Oki earthquake is the response of near-plate boundary, and the other type that of the far-field plate boundary.

The short-term variations subtracted the secular trends from the correlated seismicity rates fluctuations also display the subannual to tri-annual variations as shown in Fig. 6.60, being likely to those of the whole and SW and NE Japan described earlier. The z_2 short-term variation shows the near-annual variation in the whole periods from 1998 to 2018, and it is probably similar to those of the short-term cyclic variations appeared both in NE Japan and SW Japan. In order to check the effect of noise addition in the time series of seismicity rates, the correlated z_1 to z_4 of their data noisy data matrix is similar to those without random noise addition although it is interesting that the short-term variations of the z_1 to z_3 may be strengthened by addition of noisy fluctuations as shown in Fig. 6.69.

6.11 Partial b123 and b234 Value and Correlated Seismicity Rates

As stated earlier, the partial b123 and b234 values are controlled by the crack size populations of the inter-seismic situations in the solid earth interior. Therefore, it may be suggested that the correlated seismicity rates z_i are probably governed by the above partial b-values. Thus, it should be surveyed that the relations between the correlated seismicity rate time series and the correlated partial b-values time series

Fig. 6.70 Relations of the correlated partial b-value changes z_1 (left) and z_2 (right) of b234 and b123 and the correlated seismicity rates z_1 and z_2 of the crust in the Japanese islands region, showing the clusters of A, B, C, and D

(Fig. 6.44) can be observed in the diagram of two correlated time series. In Fig. 6.70, it seems that the z_1 and z_2 of the crust have positive correlations with those of the partial b-values b_{234} with similar locality cells contribution intensity between them. For the apparent anti-phase relation between z_{2s} of the crust and the b123 is resulted from the negative direction of the eigenvector of the b-value covariance matrix, it is consistent with that of the b234. Therefore, it concludes that the correlated seismicity rates increase roughly with increasing the partial b-value.

In this study, the partial b-value time series of the locality cells are not the case, but their correlated b-value time series are considered to extract the common features of their fluctuations. Furthermore, the largely correlated seismicity rates suggest the occurrence of the large earthquakes associated with large post-seismic activity. It remains, however, as a problem that the correlated partial b-value change is responsible for the regional differential stress change or the state change of the crack size population as noted in this study. The state change of crack size population may be largely due to the change in volume fraction of the fluid phase filling open cracks. Thus, the effective stress by water accumulation in the crack should change sensitively and propagate continuously. And thus it may result in the partial b-value lowering that means the increase of large cracks (both shear crack and open crack). Therefore, the fluid flux of water seems to be the reason of the partial b-value change associated with large correlated seismicity rates.

6.12 Correlated Seismicity Between the Global and Japanese Islands Region

In the preceding chapters, the author has discussed the characteristic features of correlated seismicity rates of the global plate boundary zones and the Japanese islands region in order to look at the cyclic variations and the general secular trends, and short-term variations. The mechanics of the solid earth is surely controlled by the plate motion and the plume dynamics in the long-term scale; but in the case of short term, it is followed by the various processes involving elastic and plastic deformation.

As a result, it may be concluded that the time series of correlated seismicity rates among many locality cells set by given volume along the plate boundary zones in the world and in the Japanese islands region have the secular trends of periodical variations with various frequencies.

In the global seismicity, because of the small to intermediate magnitude earthquakes studied in this study, the short-term cyclic fluctuations with subannual to tri-annual variations are not obvious in the time series of the correlated seismicity. In contrast, the short-term periodic fluctuations of near-annual variations are possibly found both by micro-seismicity variation and by partial b-value that is called b_{123} and b_{234} values in the downscaling Japanese islands region. The reason of these different short-term variations between the global seismicity and local seismicity rates is considered to be due to two cases: One is the difference of the magnitude size used here and the other is due to the difference of the mechanics controlling the global and local seismicity rates along the plate boundary. Therefore, it should be required to take a comparison between the global and local correlated seismicity rates. In the next sections, therefore, the direct comparison between their correlated seismicity rate time series is investigated as follows.

The time series of the correlated seismicity rates of the global plate boundary zones are displayed again together with those of the Japanese islands regions in Fig. 6.71. The time series of them are apparently coherent with each other: The sharp synchronous change between the global and Japanese trends is seen on 2011 but on about 2008–2010 coeval change with some phase difference appears although the synchronous time series of them are unclear at the time of 2004 Sumatra giant earthquake.

On the other hand, the synchronous relations of their time series are possibly inferred in more detail by means of the phase map of the z_1s of the global and the Japanese regions as shown in Fig. 6.72. It is clear that the four clusters can be identified, and they change with advancing the time. Moreover, the z_1s of them in a cluster move toward another second cluster in the period from 1998 to 2008 and then jump slightly into the second cluster. These z_1s return to the first cluster, and then they jump largely to the transient state at the high z_1s region of the third cluster. The third and fourth clusters after 2011 are referred to be the newly coming seismicity phases. Furthermore, the correlated seismicity rates of the global subduction zones appear nearly anti-phase behavior in the large z_1 levels as is seen in Fig. 6.71.

On the other hand, the z_2s of the global and the Japanese regions behave as mutually coherent relationship as shown in Fig. 6.71: The secular trend of the global correlated seismicity rates increases from 1998 to 2011 and decreases until 2016 judging from the basis vector orientation is negative in the case of z_2 of the global correlated seismicity, on the other hand, that of the Japanese regions increases at that time and then decreases until 2018. As noted in the z_1s phase map, the z_2s relationships between the global and the Japanese regions are also manifested by the phase map of them as seen in Fig. 6.73. From these figures, it seems that the coherent relationship between z_2s of the global and the Japanese regions during 1998–2011 before Tohoku-Oki earthquake and after 2011 two patterns of large z_2 and small z_2 domains of the Japanese correlated seismicity rates can be found in the figures. The

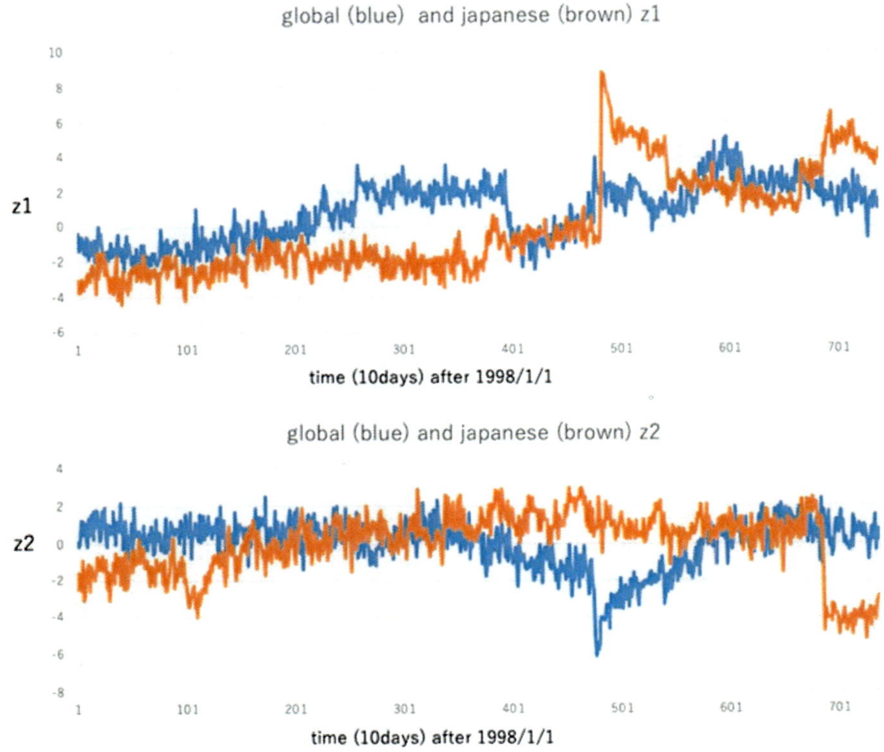

Fig. 6.71 Comparison of the correlated seismicity rates z_1 and z_2 between the global subduction zone (M4-M5) and Japanese (M1–M2) region

Fig. 6.72 Diagrams of the correlated seismicity rates z_1 of the global subduction zone and Japanese region, showing four clusters A, B, C, and D. The right-hand figure is the average z diagram

Fig. 6.73 Diagrams of the correlated seismicity rates z_2 of the global subduction zone and Japanese region, showing one cluster and two spines. The right-hand figure is the average z diagram

former is probably related to the post-Tohoku-Oki giant earthquake but the reason of the latter is not clear. It is noteworthy that scares signal at the 2004 Sumatra giant earthquake can be seen in the phase map of 1998–2011 as a small cluster at small z_2 of the global. It looks likely the signals of cluster in large variations of the z_2 of the global as shown previous figure.

According to Okada et al. (2017), the characteristic variations of global correlated seismicity rates are strongly influenced by the seismic activity of the NE Japan district, and thus, the z_1s phase map described previously shows the strong correlation between the global and the Japanese regions during 2011–2013 as shown in Figs. 6.71 and 6.72. This is the manifestation of the strong correlation of the seismic activity in the global and the downscaling regions through the combined seismic activity of the globally correlated seismic regions such as Chile, Tonga-Kermadec, and Sumatra. In addition, the seismic activity of the global ridge systems is also correlated with that of the global subduction systems, and thus, it is mechanically combined with downscaling Japanese regions too as discussed in the previous chapters of the global seismic activity.

Judging from the fact that the strongly correlated seismicity rates in the global subduction and ridge boundaries are nearly synchronized with each other; the down-scaling correlated seismicity rates of Japanese islands region are controlled by the motions of surrounding Pacific and Philippine Sea plates in terms of the seismic activity of Tonga-Kermadec, Chile, and Alaska plate boundaries. These interactions between the global and the regional mechanics in the surficial solid earth are mainly governed by the global plate motions and the difference of them buffered by the deep convective motion of the whole mantle. Therefore, the modeling of the plate boundary seismicity should be required of the viscous and frictional stresses operating along on the plate boundaries and between the plates and underneath mantle as discussed in the later sections.

References

Anderson OL, Grew PC (1977) Stress corrosion theory of crack propagation with application to geophysics. Rev Geophys Space Phys 15:77–104

Bishop CM (2006) Pattern recognition and machine learning. Springer, New York

Fryer P, Ambos EL, Hussong DM (1985) Origin and emplacement of Mariana forearc seamounts. Geology 13:774–777

Fukao Y, Ohbayshi M, Inoue H, Nendal M (1992) Subducting slabs stagnant in the mantle transition zone. J Geophys Res 97:4809–4822

Heki K (2007) Secular, transient and seasonal crustal movements in Japan from a dense GPS array: implicate for plate dynamics in convergent boundaries. In: Dison T, Moore C (eds) Seismogenic zone of subduction thrust faults, pp 512–539

Heki K (2010) Co-seismic gravity changes of the 2010 earthquake in central Chile from satellite gravimetry. Geophys Res Lett 37:24306. https://doi.org/10.1029/2010GL045335

Ide S, Shelly DR (2007) A scaling law for slow earthquakes. Nature 447:76079. https://doi.org/10.1038/nature05780

Iio Y, Sibson R, Takeshita T, Sagiya T, Shibakusa B, Nakajima TJ (2018) Crustal dynamics: unified understanding of geoynamic processes at different time and length scales. Earth Planets Space 70:97

Kimura J-I (2017) Modeling chemical geodynamics of subduction zones using the arc basalt simulator version 5. Geosphere 13(4):992–1025. https://doi.org/10.1130/GES01468.1

Matsuo K, Heki K (2011) Coseismic gravity changes of the 2011 Tohoku-Oki earthquake from satellite gravimetry. Geophys Res Lett 38:7

Nakata R, Suda N, Tsuruoka H (2008) Non-volcanic tremor resulting from the combined effect of earth tides and slow slip events. Nat Geosci 1:676–678

Obara K (2002) Nonvolcanic deep tremor associated with subduction in southwest Japan. Science 296:1679–1681. https://doi.org/10.1126/science1070378

Ogata Y (1988) Statistical models for earthquake occurrences and residual analysis for point processes. J Am Stat Assoc 83:9–27

Ohnaka M, Matsu'ura M (2002) The physics of earthquake generation. University of Tokyo Press, 378 pp (in Japanese)

Okada A, Toriumi M, Kaneda T (2017) Spatial and temporal pattern of global seismicity extracted by dimensionality reduction. Int J Geol 11:26–34

Omori S, Kamiya S, Maruyama S, Zhao D (2002) Morphology of the intraslab seismic zone and devolatilization phase equilibria of the subducting slab peridotite. Bull Earthq Res Inst Univ Tokyo 76:455–478

Schwartz S, Rokosky JM (2006) Slow slip events and seismic tremor at circum-Pacific subduction zones. Rev Geophys 45:RG3004. https://doi.org/10.1029/2006rg000208 (2007)

Shelly DR, Beroza GC, Ide S (2007) Non-volcanic tremor and low-frequency earthquake swarms. Nature 446:305–307

Sugimura A (1960) Zonal arrangement of some geophysical and petrological features in Japan and its environs. J Fac Sci Univ Tokyo Sect II 12:133–153

Tanaka S (2012) Tidal triggering of earthquakes prior to the 2011 Tohoku-Oki earthquake (Mw 9.1). Geophys Res Lett 39:L00G26. https://doi.org/10.1029/2012gl051179

Tatsumi Y (1986) Formation of the volcanic front in subduction zones. Geophys Res Lett 13:717–720

Toriumi M (2009) Principal component analyses of seismic activity in the plate boundary zone of northeast Japan arc. J Disaster Res 4(2):209–213

Chapter 7
Correlated Seismicity of the Northern California Region

Abstract In the Northern California region, the transform fault and related transcurrent faults are running between oceanic ridges, and this tectonic situation is different with that of the subduction boundary in the Japanese region. The characteristic features of the seismicity in the Northern California region are possibly investigated by the correlated seismicity likely to the Japanese seismicity.

Keywords Characteristic features of correlated seismicity in Northern California · Partial *b*-value variation · Relation in correlated seismicity between global and Northern California

7.1 Introduction

In order to clarify the downscaling seismicity of the Northern California, the huge amounts of small earthquakes are taken from the databases of NCDSE in USGS and Berkeley of Univ. of California, and these are stored in new databases of ESM. The structural geology of the Northern California is controlled by the large transcurrent faults such as San Andreas Fault and Heyward Fault. These large faults belong to on-land transform faults system between the Pacific plate and the North American plate, and they are terminated to the triple junction of the Cascadia subduction boundary and incipient Juan de Fuca–Pacific plate spreading ridge. Furthermore, there are two geologically active areas found: One is the Long Valley Caldera and the other is the Geyser Geothermal Field (Norris and Webb 1990). These two regions are very active in the micro-seismicity in the crust due to the circulation of hydrothermal solution in the crust and magmatic activity under the Caldera.

There are many studies of the seismic activity in the Northern and Southern California regions because of large active earthquakes often occurred near San Francisco.

Electronic supplementary material The online version of this chapter (https://doi.org/10.1007/978-981-15-5109-3_7) contains supplementary material, which is available to authorized users.

M. Toriumi, *Global Seismicity Dynamics and Data-Driven Science*, Advances in Geological Science,
https://doi.org/10.1007/978-981-15-5109-3_7

In addition, there are abundant seismic signals around the Cascade Mountain Range such as Mt. Saint Helene and Mt. Shasta. The seismic activities are in the regions of geothermal fields of Geyser and Long Valley Crater. However, the central California region is not the subduction zone in contrast with the Japanese islands region. Therefore, it needs to compare the mode of difference between the subduction zone and transform zone in terms of the seismicity rate of the plate boundary (Waldhauser and Schaff 2008).

7.2 Seismicity Cloud of the Northern California Region

To be easily visible in seismicity data, the seismicity rates in the locality cells mapped in Fig. 7.1 of Long Valley Crater (LVC), Geyser Geothermal Region (GGR), San Francisco area (SFA), and others are chosen for time series diagram as shown in Fig. 7.2 and also shown in 3D diagram of their logarithmic values (Fig. 7.3). It is obvious that the large spine develops from the central cloud toward the LVC and small spine does toward the GGF orientation as seen in Fig. 7.3, suggesting the dominant seismic activity of small-to-micro-earthquakes responsible for volcanic and hydrothermal activity in the crust. On the other hand, the time series of seismicity rates of locality cells can be classified into two types as seen in Fig. 7.2: One is the continuously active time series, and the other shows the several sharp spikes with weak continuous seismicity. The latter type is appeared in the LVC and related large fault, but the former one is along the Heyward Fault zone and GGF. It is probable that the geothermal fluid activity and related fault activity as seen in seismic activity are relatively continuous, but the magmatic activity centered in the LVC is associated with episodic fault activity in the Northern California region as shown in Fig. 7.2. The data cloud of the Northern California region is shown in Fig. 7.3, suggesting the broad spine extending from the core cloud.

The correlated seismicity rates should be strongly responsible for these two types of seismic activity along with the large faults and centered source of seismicity near the geothermal and magmatic regions. Thus, the following coordinate transformation was carried out by means of deterministic PCA (SVD method) and sparse-modeled PCA with LASSO as like as the data analyses of Japanese islands region and global earth.

7.3 Correlated Seismicity Rates in Northern California

As shown in Fig. 7.4, the time series of the correlated seismicity rates z_1 to z_4 obtained by deterministic PCA and sparse PCA methods are identical patterns: The time series of z_1 appears the secular trend with long periodical change and that of the z_2 does periodical change with about 10 years and near annual cycles as shown in Southern California region by Johnson et al. (2017). The feature patterns of z_3 and z_4 display unclear periodic variations with about several years of cycles, but these

Fig. 7.1 Map showing the locality cells by the number (1–96) in the Northern California region. Reference localities of LVC, GGR, SFA are Long Valley Crater, Geyser Geothermal Region, and San Francisco area, respectively

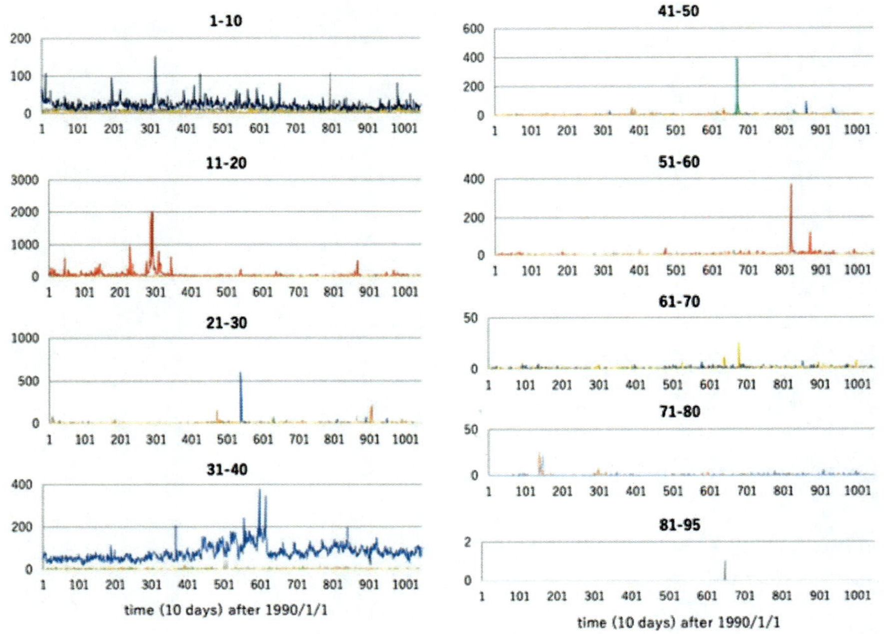

Fig. 7.2 Time series of the seismic activity in the locality cells of the Northern California region from 1990 to 2019. Cell numbers are shown by the top figure

are very noisy. The variances of the z_1 and z_2 fluctuate through 1990 to 2019, but the fluctuation patterns of the early phase (1990–2004) are different from those of the later phase (2005–2019) as shown in Fig. 7.5. In the diagram of variances of z_1 and z_2 in Fig. 7.6, there makes a massive cluster around (0.4, 0.6) in the early phase but does scattered patterns extending 1.5 of z_1 and z_2, suggesting that the fluctuation of the correlated seismicity becomes very large in the later phase.

The z_1 and z_2 migration averages of time series show the cyclic patterns with various frequencies: The periodicity of the z_1 time series is about 20 years, but that of the z_2 one displays about 10 years accompanied with 1–2 years cycles as shown in Fig. 7.7. The subsidiary cycles near 1–2 years are clearly found in the z_2 time series. On the other hand, the feature patterns of the z_3 and z_4 time series are of the noisy time series likely to Gaussian types, but they have sometimes spikes with short duration suggesting the rather large earthquakes. The Gaussian-type noisy patterns are possibly found by the diagrams of z_3 and z_4, and z_4 and z_5 signals (Fig. 7.8).

In order to understand the difference of combined patterns of the correlated seismicity rates time series, the diagram of the z_1 to z_2 is shown in Fig. 7.9, thereby suggesting that there appear two wings of their data clusters: One is the left cluster showing low level of the z_1 and the other the right wing showing the high level of the z_1. Both wings of the coupled z_1 and z_2 are dispersed by the cyclic oscillation of the z_2 components. It is important that the left wing of the z_1–z_2 relations belongs to

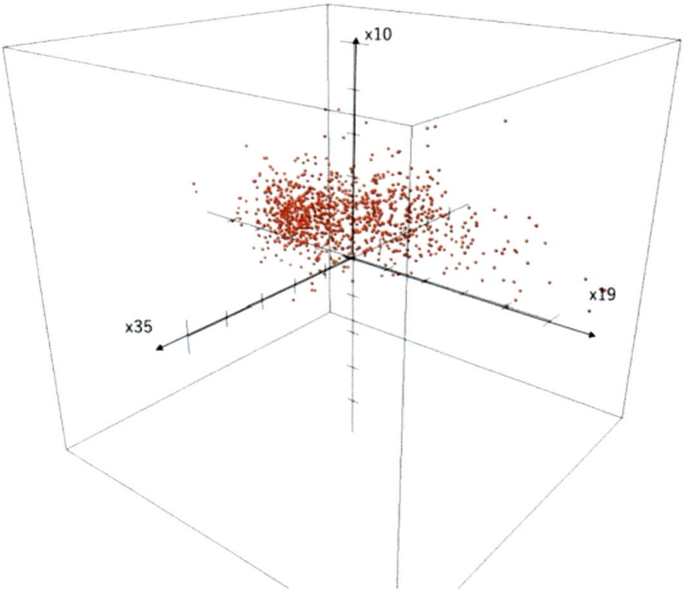

Fig. 7.3 Seismicity rates cloud of the Northern California region in the activity of three locality cells

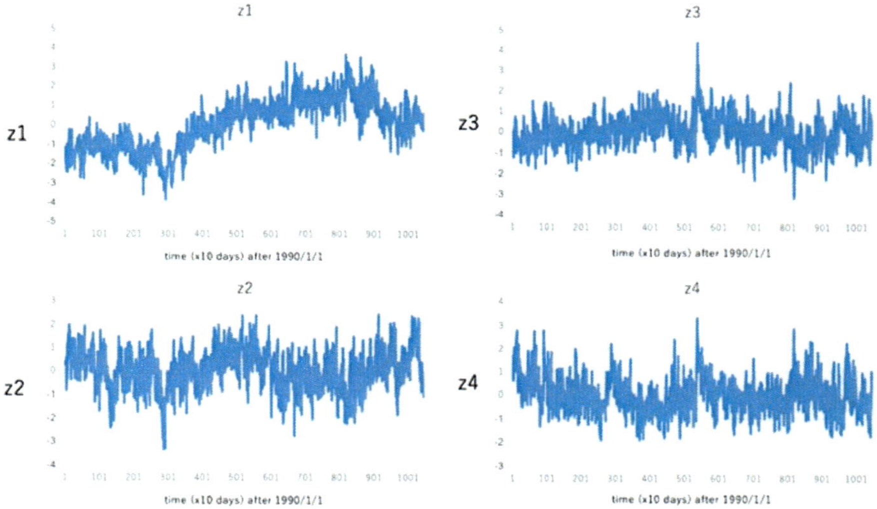

Fig. 7.4 Time series of the correlated seismicity rates z_1 to z_4 of the crust in the Northern California region from 1990 to 2019

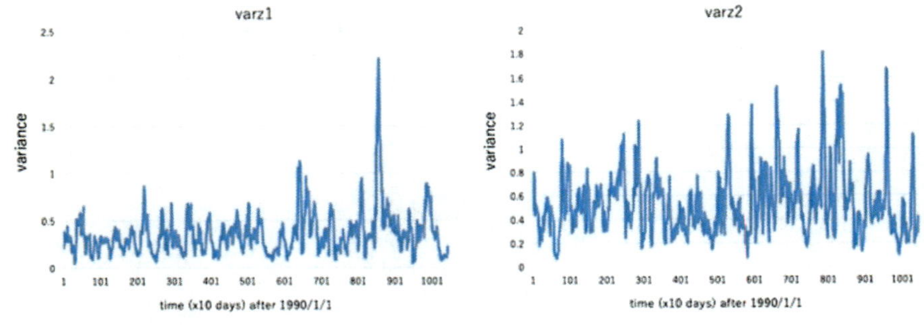

Fig. 7.5 Time series of the variance in the correlated seismicity rates z_1 and z_2 in the Northern California region from 1990 to 2019

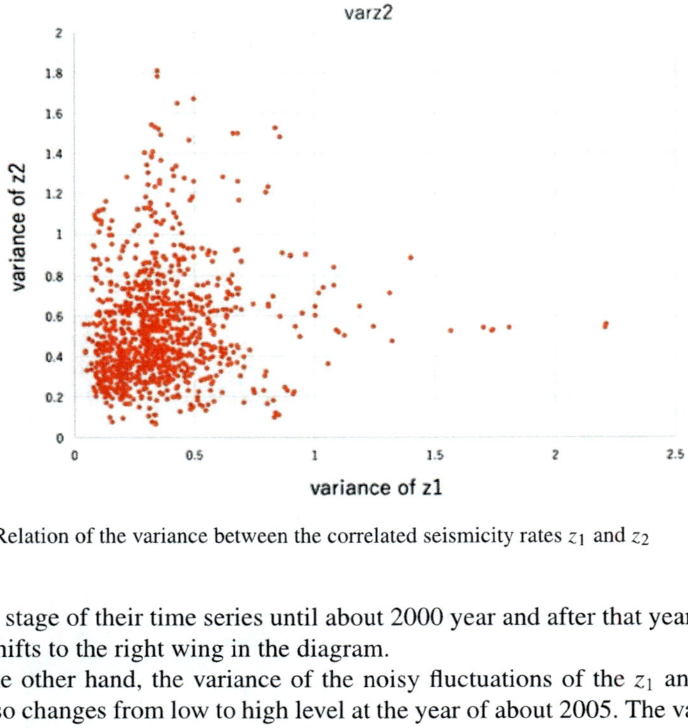

Fig. 7.6 Relation of the variance between the correlated seismicity rates z_1 and z_2

the early stage of their time series until about 2000 year and after that year, the data cluster shifts to the right wing in the diagram.

On the other hand, the variance of the noisy fluctuations of the z_1 and z_2 time series also changes from low to high level at the year of about 2005. The variance of both z_1 and z_2 after that year varies in the wide range as shown in the time series of z_1 and z_2 variances together with the diagram of their variances as shown in Figs. 7.5 and 7.6. In the periods from 1990 to 2005, the data of z_1 and z_2 variances make a simple cluster, but after 2005 they are widely scattered in the region extending in the large variances of z_1 and z_2.

It is noted that the annual variations subtracting secular trends of the z_1 and z_2 display the obviously cyclic patterns as shown in Fig. 7.10. The periodic cycles of

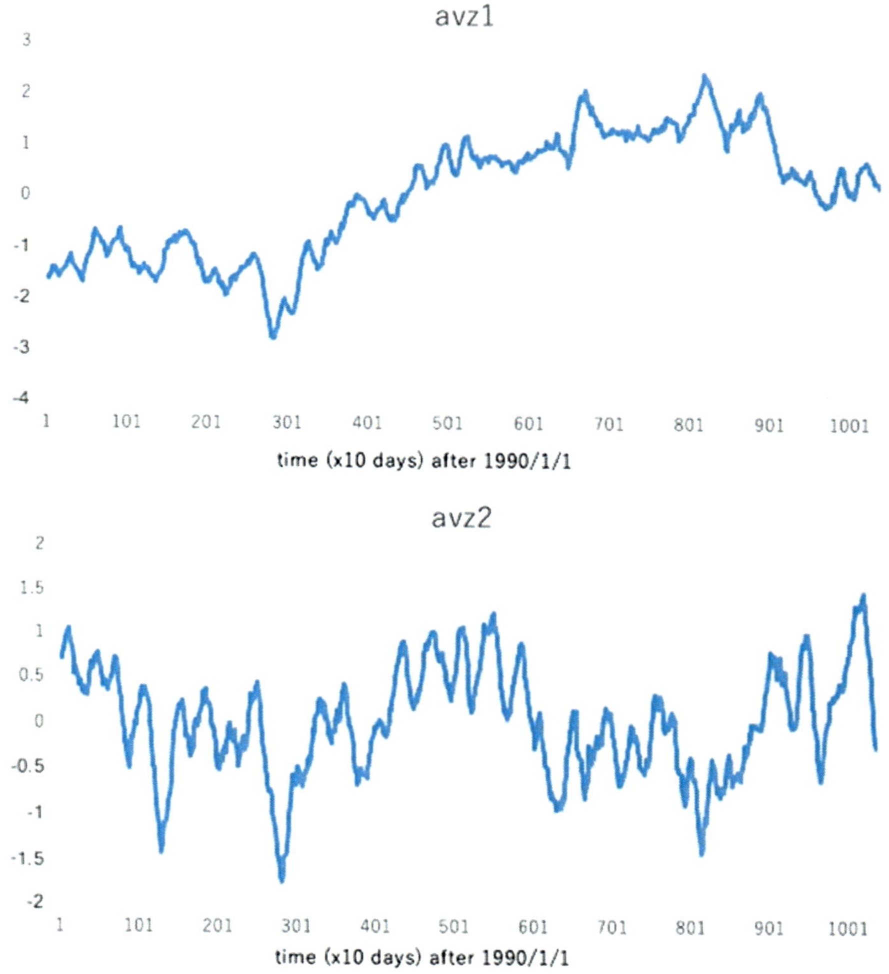

Fig. 7.7 Time series of the average correlated seismicity rates z_1 and z_2 in the Northern California region

them seem to be about 1 to 2 years at the peak of 1.5 years, as shown in the histogram of the FFT spectra of the time series of annual variation of z_1 and z_2 (Fig. 7.11). The cyclic change of z_2 is more clear rather than that of the z_1. The synchronization between their annual variations of z_1 and z_2 is not obvious from the diagram of short term z_1 and z_2 (Fig. 7.10). In this figure, there are synchronous and anti-synchronous patterns of the annual variations and scattered patterns of them, judging from the dense trends of the cluster. Considering that the z_1 correlated seismicity rates represent the coherent activity of the eastern fault system connecting the Long Valley Crater (LVC) but the z_2 does the coherent one of the western fault system

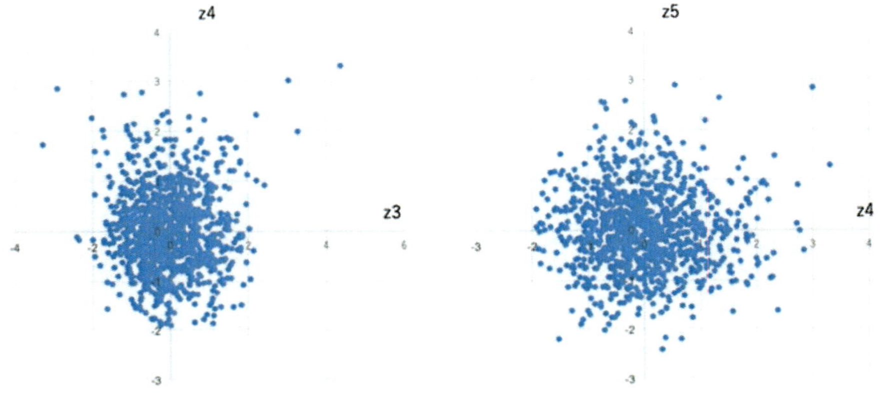

Fig. 7.8 Diagrams of the correlated seismicity rates z_3 and z_4, and z_4 and z_5 showing the noisy patterns

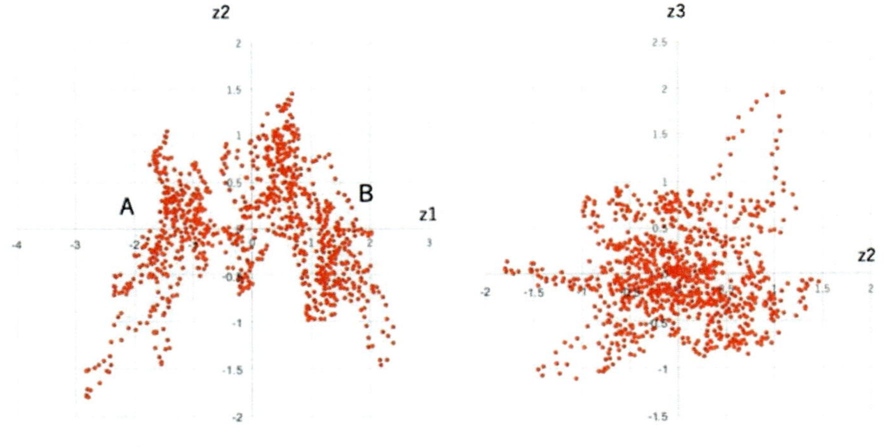

Fig. 7.9 Diagram of the relation between the average correlated seismicity rates z_1 and z_2, and z_2 and z_3 in the Northern California region from 1990 to 2019, showing the two wing-like clusters *A* and *B*

connecting Heyward Faults and Geyser Geothermal Field (GGF) as seen in the figures of eigenvector components (Fig. 7.12) as like as results of Tiampo et al. (2002), it seems that there are different switching modes of seismic activity between the eastern and western fault systems: Those are coherently synchronous and anti-synchronous and indifferent modes of the seismicity correlation. Noisy fluctuations of the z_4 time series can be seen in the both western and eastern fault zones. The pattern of the z_3 is governed by the coherent seismicity along the eastern fault system likely to that of the z_1.

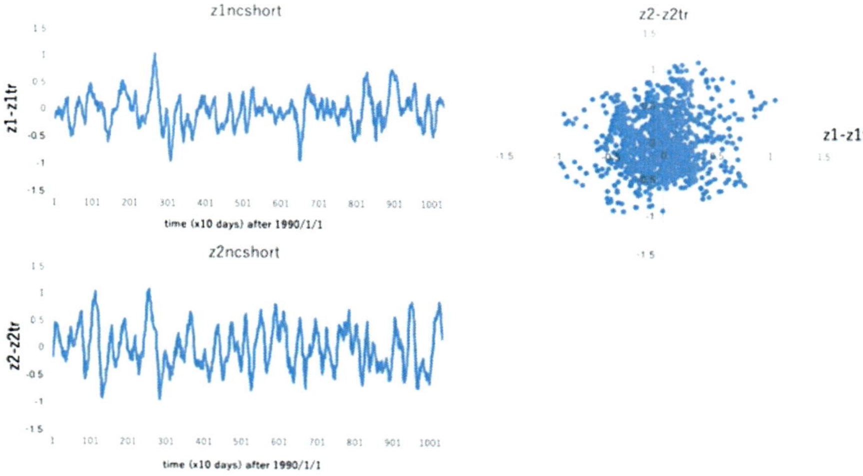

Fig. 7.10 Time series of the short-term variations subtracted by the secular trends of the correlated seismicity rates z_1 and z_2 (left) and the z_1–z_2 diagram (right) of the Northern California region

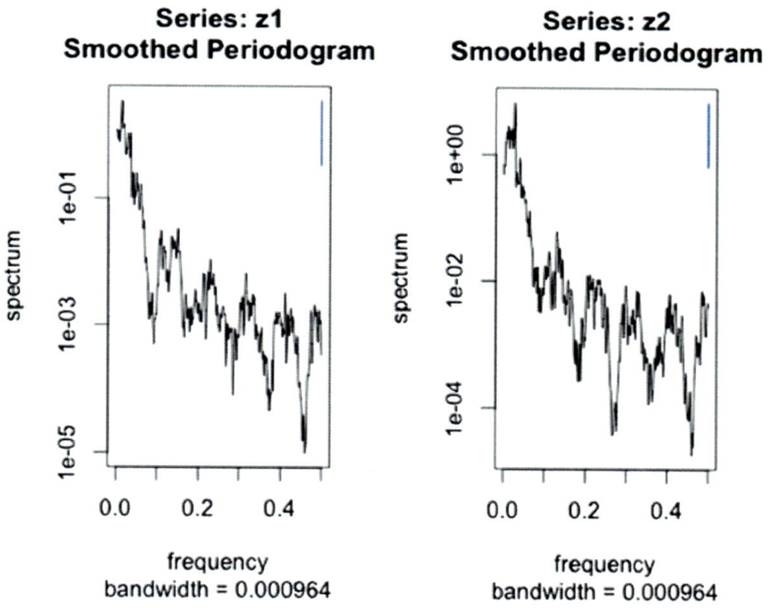

Fig. 7.11 Fast Fourier diagrams of the short-term variation of the correlated seismicity rates z_1 and z_2 in the Northern California region

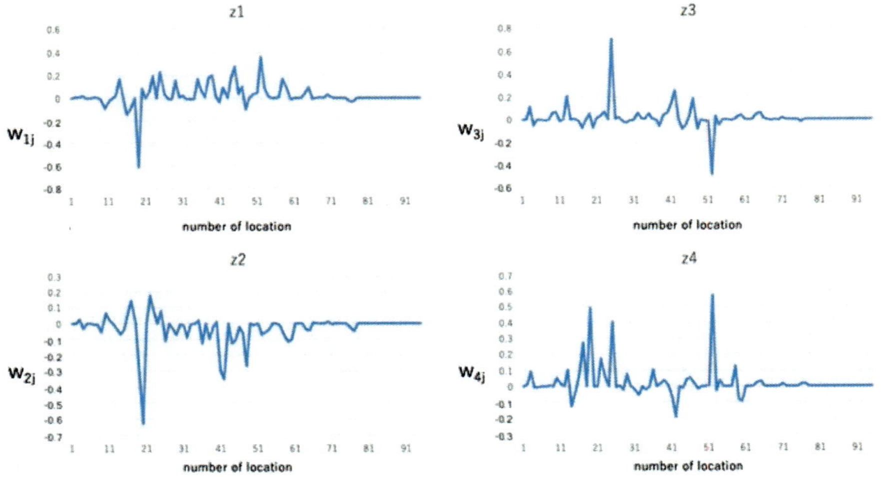

Fig. 7.12 Linear patterns of the correlation intensity (w_{ij}) of the locality cells shown in Fig. 7.1 of the crust in the Northern California region

In addition, the correlated seismicity rates of magnitude 2–3 are also investigated as same as the previous method above mentioned. The correlated time series of z_1 to z_4 of the seismicity rates of M2-3 appears in the Fig. 7.13, suggesting that the early stage of both time series of the z_1 and z_2 shows a gradual change from low to high level, but in the later stage, the patterns of them look like reverse relationship of z_1 and z_2. The diagrams of z_1 and z_2, and z_2 and z_3 are shown in Fig. 7.14, suggesting that there are two clusters in the former: One is compact but another scattered.

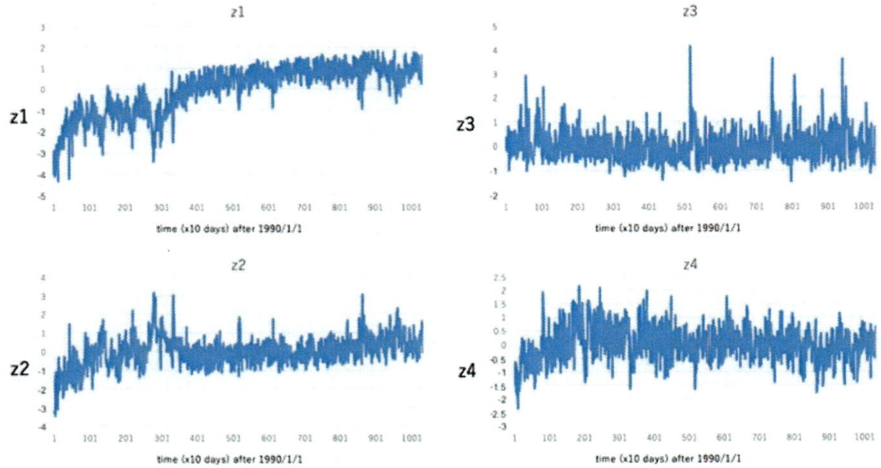

Fig. 7.13 Time series of the correlated seismicity rates z_1 to z_4 of M2-3 in the crust by sparse modeling PCA method in the Northern California region from 1990 to 2019

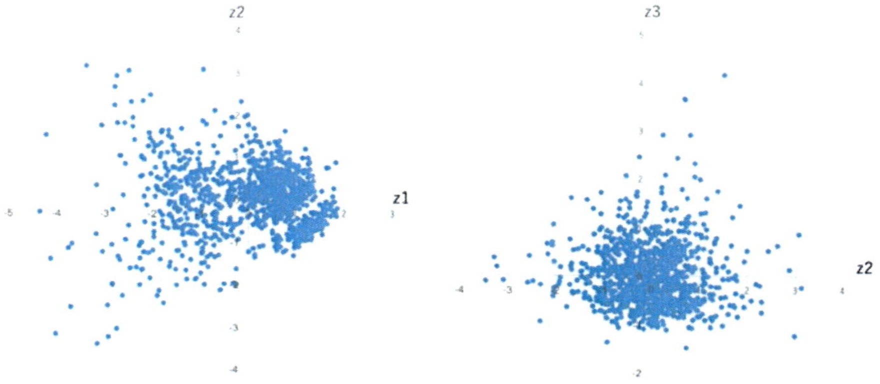

Fig. 7.14 Relation of the correlated seismicity rates z_1 and z_2, and z_2 and z_3 of M2-3 in the Northern California region

7.4 Partial *b*-Value Variations of the Northern California Region

The partial b-value, b123 fluctuations can be also investigated by means of correlated patterns of the time series as same as in the case of Japanese islands and global seismicity. As shown in Fig. 7.15, the correlated partial b-value variations display the gradual change from 1990 to 2019 in the time series of the z_1 of b-value, but that of z_2 looks like the noisy patterns of Gaussian type as shown in Fig. 7.16. In the case of the Japanese islands regions, the annual variations of partial b-value changes are clear in the correlated patterns; however, those in the case of the northern California appear only incipient. The migration average patterns of them become rather clear, and they display the periodic variation having 1–2 years cycle (Fig. 7.17). The histogram of the peak to peak of the correlated seismicity rate time series displays the representative periodicity about 300–350 days as shown in Fig. 7.18. The synchronous and non-synchronous variations between z_1 and z_2 of the correlated partial b-value change are shown by the diagram of the z_1 and z_2 in Fig. 7.18. Many of data points indicate the concentration at the center and scattering around the center, indicating the noisy correlations between them. It is noted that the similarity in annual variation between their regions may be due to difference in the structural framework of the geology: Japanese islands region is the subduction boundary zones, but the Northern California

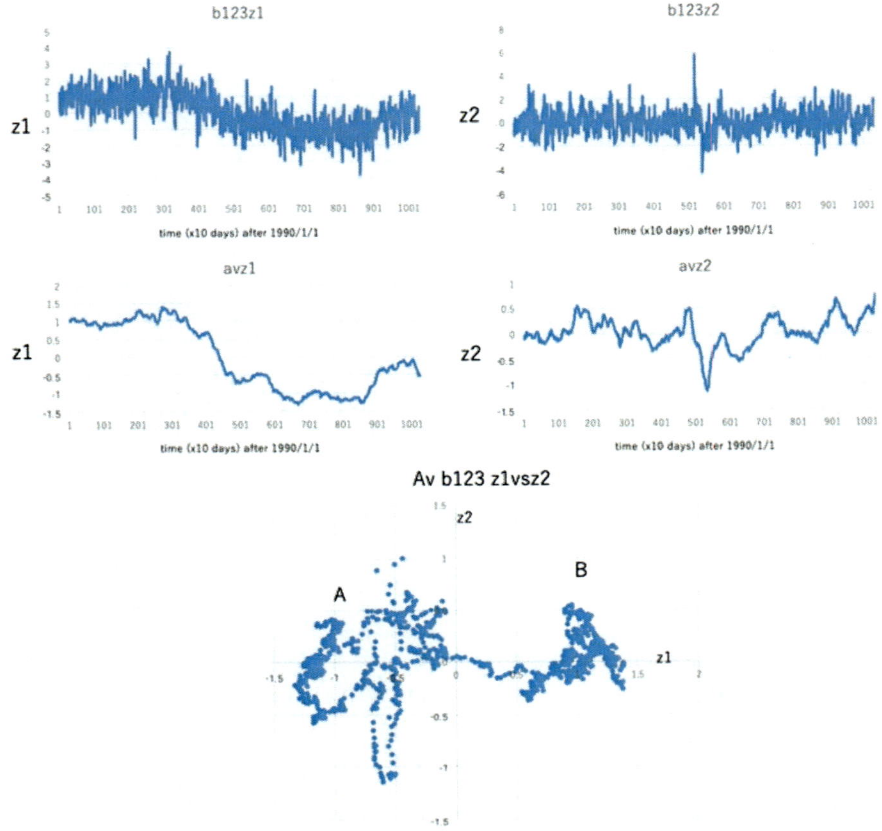

Fig. 7.15 Time series of the correlated *b*-value change z_1 and z_2 of b123 and their average change (middle) and their z_1-z_2 diagram (bottom) showing two wings *A* and *B* of the Northern California region

region is the transform boundary region with large geothermal area and giant Caldera region. Thus, it needs to go back to the partial *b*-value changes in the local seismicity of the Northern California to compare the tectonic effects of geothermal and volcanic activities.

The partial *b*-value changes of the many localities in the Northern California region are possibly classified into two types: One is rather cyclic with small amplitude, and the other varies largely with incipient periodicity as shown in Fig. 7.15. The former appears in the fault zone, but the latter is characterized in the LVC and GGF regions, suggesting the large amplitude of the partial *b*-value change is probably due to the magmatic and hydrothermal activities in the crust.

The partial *b*-value change will be discussed in Chap. 10, and it represents the variation in the order parameter, *s* as follows;

Fig. 7.16 Relation of the z_1 and z_2, and z_2 and z_3 of the correlated b-value (b123) change of the Northern California region

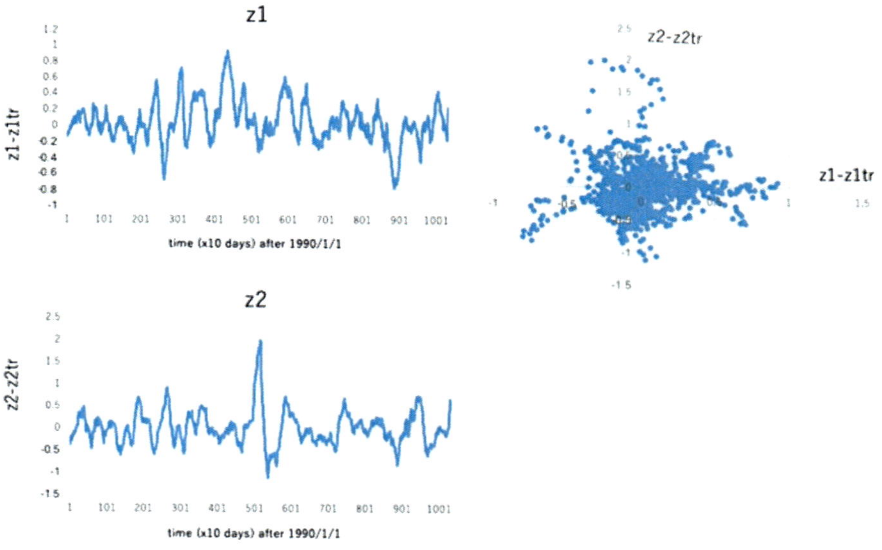

Fig. 7.17 Time series of the short-term variations of the correlated b-value change z_1 and z_2 for b123 and the relation between them in the Northern California region

$$s = \exp(-b) \tag{7.1}$$

Therefore, the annual change of partial b-value means the variation of the order parameter defined by large crack percolation against the small crack population. Furthermore, it is noteworthy that above partial b-value means the probability density

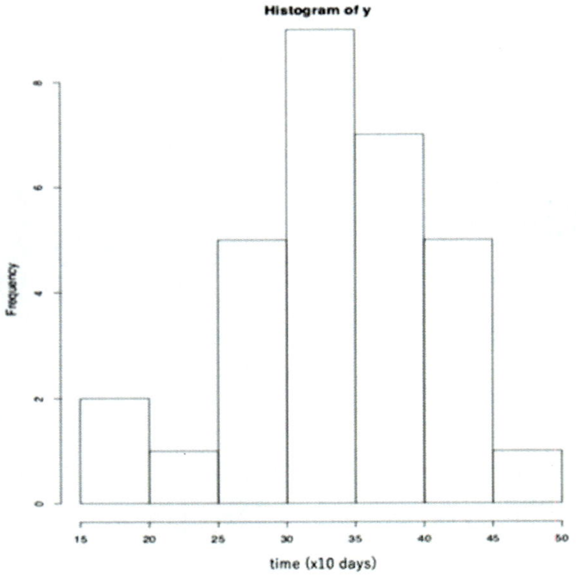

Fig. 7.18 Frequency histogram of the peak-to-peak time in the short-term variation z_1 of the correlated *b*-value change b123 in the Northern California region

ratio of the small shear crack against large one, and thus, it seems that the partial *b*-value is a potential indicator of the correlation magnitude between these shear cracks as well as the order parameter s as shown in Fig. 7.19. In other words, though the correlated partial *b*-value, *z* represents its spatial and temporal correlation degree,

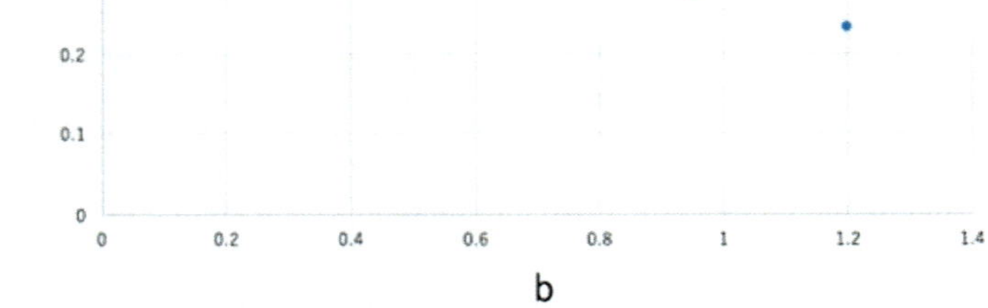

Fig. 7.19 Model relationship between the partial *b*-value of the micro-seismicity *b* and the order parameter *s* (see text)

the partial b-value itself does the correlation degree of probability in size of shear cracks in the given locality. Therefore, it seems that the correlated partial b-value change should mean the temporal changes of regional variation of the shear crack size probability ratio controlled by the mechanics of the shear crack growth.

By the way, the small shear crack growth in the subcritical conditions is governed by differential stress, temperature, fluid pressure of water, and the population density of the shear crack increases with increasing porosity and water content. For the shear crack may be associated with open crack through jog and kink formation of cross slip, the porosity and water content of the rock mass should increase with increasing shear crack population/density. Therefore, the fluid influx into the rock mass from the dehydrated subduction plate or the earth surface requires for the increase of shear cracks density. It leads that when the fault slippage such as SSE is active and causes increasing the pores along the shear faults, the partial b-value and also its correlated b-value should change from low to high level. The hydrothermal and magmatic activity must also increase the pores filled with fluid, and then, it increases partial b-value and correlated partial b-value.

The z_1 component of correlated partial b-value decreases gradually from 1990 to 2015 and increases in turn until 2019. The main constituent localities of this z_1 component are lined along the fault zone extending from the Long Valley Crater. Therefore, it concludes that the b-value change along the LVC-related fault zone activity is probably due to the dynamics of high-temperature fluid flux and related shear crack population increase as noted previously. Then, it seems to be suggested that the correlated seismicity rates, z_1 along the fault zone mentioned above should manifest the similar trend that shows gradual change from 1990 to 2019 as shown in Figs. 7.4 and 7.12, being likely to the variation of z_1 of partial b-value.

On the other hand, the z_2 of the partial b-value has the main constituents of the locality cells limited in the vicinity of LVC, and its variation shows a little cyclic time series as shown in Figs. 7.4 and 7.12 in contrast with the z_1 pattern. However, the z_2 component of the correlated seismicity rates has main constituents of the locality cells belonging to both LVC and GGF. The feature patterns of the z_2 of the correlated seismicity rates appear the annual variation (Figs. 7.7 and 7.10) and longer cyclic change having about decade, suggesting that there is something like synchronous activity between the fault zones from LVC and GGF.

7.5 Comparison Between Global Subduction Zones and Northern California Region

As noted in the previous sections, there are three types of plate boundary zones: The one is the subduction zone, the second is the mid-oceanic ridge zone, and the third is the transform fault zone. In this section, comparison of the correlated seismicity rates between the global subduction zone and the Northern California transform fault zone will be discussed. First let us look at the relations between their time series: In the

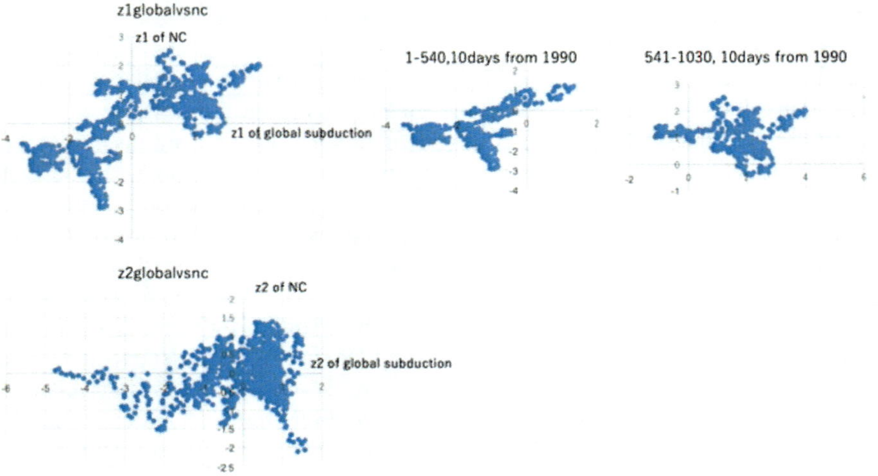

Fig. 7.20 Diagram of the correlated seismicity rates of the global subduction zone and the Northern California region, and their successive change with time (right)

diagram of the correlated seismicity rates z_1 between the global subduction zone and Northern Californian region (Fig. 7.20), it is obvious that the synchronous relations between them appear in the general trend from 1990 to 2018, but in some periods of 1990–2005 and 2015–2017, non-synchronous relations are possibly identified. As pointed out in the sections of global seismicity patterns, the z_4 components of the correlated seismicity rates belong to the locality cell network containing the subduction zone of Mexico as discussed in Chap. 5, and thus, the diagram of the z_4 of the global subduction zone and the z_1 of the Northern California may be taken into consideration. It suggests that there is no correlation between them, but there is a Gaussian-like random noisy pattern of z_4 of the global case. Therefore, it may conclude that the seismicity dynamics of the Northern California is controlled by the main correlated seismicity dynamics of the global subduction zone related to the western Pacific region (z_1 node) but not by the network subsystem of eastern Pacific region related to z_4 of the global case.

The reason why there are synchronous and non-synchronous modes in the relation of the z_1 components between the global and the Northern California regions may be derived from the long-term variation depending on the plate motion and short-term one due to magmatic and geothermal activities. The latter seismicity may result in the local disturbance of the correlated seismicity rates.

References

Johnson CW, Fu YN, Burgmann R (2017) Seasonal water storage, stress modulation, and California seismicity. Science 356(6343):1161–1164. https://doi.org/10.1126/science.aak9547

Norris RM, Webb RW (1990) Geology of California. Wiley, NewYork, p 541

Tiampo KF, Rundle JB, McGinnis S, Gross SJ, Klein W (2002) Eigenpatterns in southern California Seismicity. J Geophys Res 107(B12):2354. https://doi.org/10.1029/2001JB000562

Waldhauser F, Schaff DP (2008) Large-scale relocation of two decades of Northern California seismicity using cross-correlation and double-difference methods. J Geophys Res 113:B08311. https://doi.org/10.1029/2007JB005479

Chapter 8
Model of Seismicity Dynamics from Data-Driven Science

Abstract The multivariate time series of the correlated seismicity rates newly obtained by sparse modeling in the global and regional systems are potential indicators of seismicity dynamics. The sets of these parameters with successive time are mapped and imaged as attractors in the phase diagram, introducing possibly the minimal dynamic equations.

Keywords Minimal dynamic model of global seismicity · Nonlinear model of correlated seismicity dynamics · Comparison of correlated displacement features in simulated circular—connected slider blocks model

8.1 Minimal Model of Global Seismicity Dynamics

First, the author intends to classify the basic trends of the global seismicity which are separated from the giant earthquake trends such as 2004 Sumatra and 2011 Tohoku-Oki earthquakes as shown in global trends of correlated seismicity rates (Fig. 8.1). Looking at the diagram of z_1 to z_3 as shown in Fig. 8.2, the main trends of them should pass three compact clusters of *A*, *B*, and *C* and the abnormal trends with scattered points from C cluster. These trends of progressive change of clusters are probably clear in the diagram of z_1 and its variance, and z_2 and its variance as shown in Fig. 8.3. It is obvious that the abnormal trend to the z_2 direction is just for the 2004 Sumatra earthquake and that to the z_3 direction is for the 2011 Tohoku-Oki earthquake, as being also shown by cluster diagrams obtained from "mclust" in R package (Fig. 8.4). In this diagram, the main path of compact clusters is surely identical.

The spectrogram of the z_1 time-series-divided six segments as shown in Fig. 8.5 can be expressed in the variation change of the locality cells intensity that means the weight coefficients of w_{ij} for *i*-th *z* components. In Fig. 8.6, it is possible to be

---

Electronic supplementary material The online version of this chapter (https://doi.org/10.1007/978-981-15-5109-3_8) contains supplementary material, which is available to authorized users.

M. Toriumi, *Global Seismicity Dynamics and Data-Driven Science*, Advances in Geological Science,
https://doi.org/10.1007/978-981-15-5109-3_8

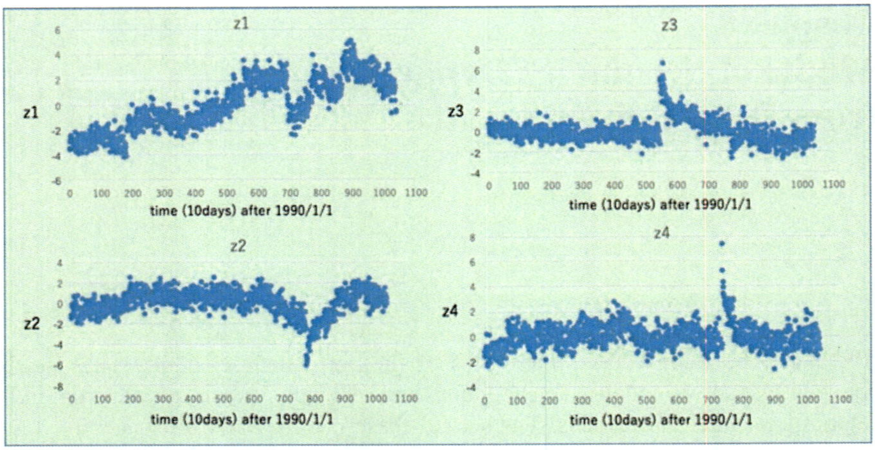

Fig. 8.1 Time series of the main correlated seismicity rates of the global subduction zones showing the secular continuous change and discontinuous rapid change. Horizontal axis is the time in 10 days unit, and vertical axis is the correlated logarithmic seismicity rates minus average one

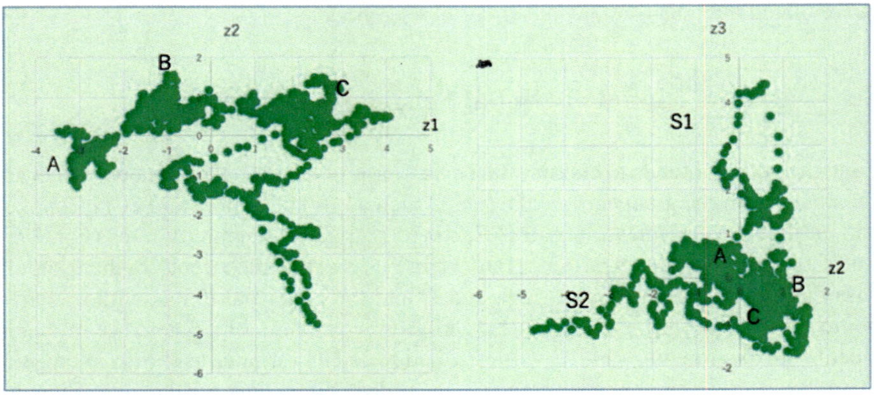

Fig. 8.2 Diagrams of the correlated seismicity rates z_1 to z_3 by averaged values in the global subduction zones. The rather discontinuous clusters in left diagram are seen. Clusters A, B, and C are the same as Fig. 5.30 and S1 and S2 are spines of 2004 and 2011, respectively

considered that before 2004, the successive trend from *A* to *B* comprises with the weight patterns of 1 and 2 of Fig. 8.5 but to the 3 of figure it changes abnormal trend, and that before 2011, the patterns of the weight coefficients change to be active in many locality cells and after this change the abnormal trend toward the z_3 component with 2011 Tohoku-Oki earthquake. After 2012, the z_1 looks like the pattern 4 with active many locality cells in Fig. 8.6. And, the situation of the present after 2011 is plotted in the cluster of C in the three z_i diagram (Fig. 8.3).

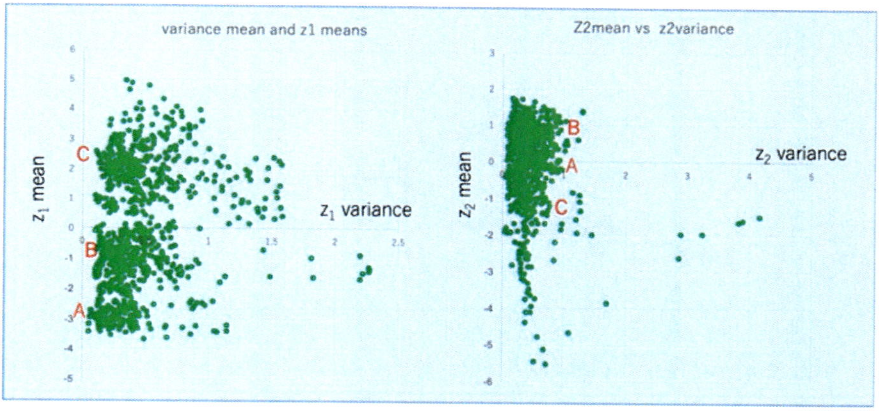

Fig. 8.3 Diagrams of the mean correlated seismicity and the variance of z_1 and z_2 in the global subduction zones. A, B, and C are the same clusters as those in Fig. 5.30

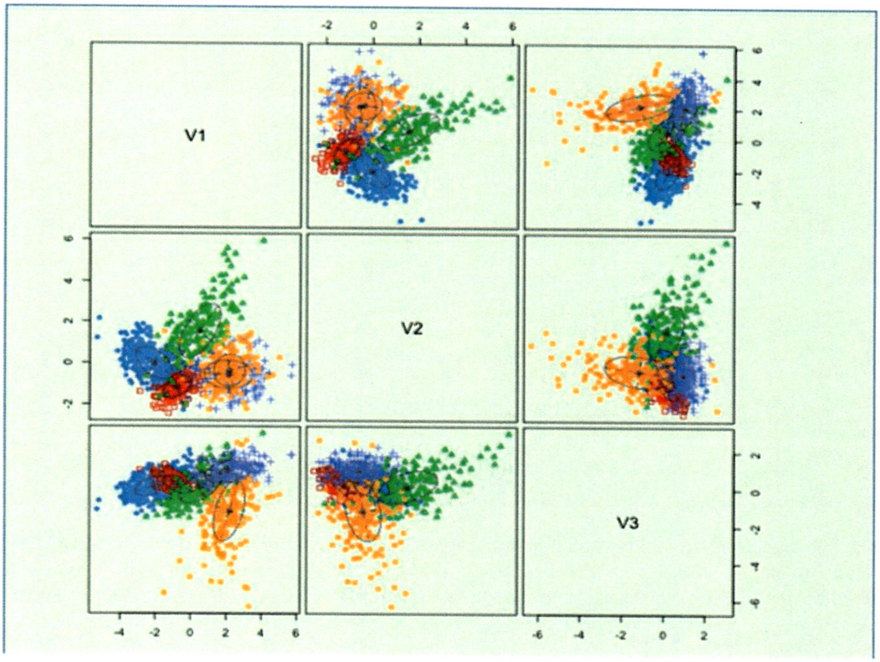

Fig. 8.4 Clusters of the correlated seismicity rates and spines of 2004 (brown) and 2011(green) in the z_1(V1), z_2(V2), and z_3(V3) in the global subduction zones

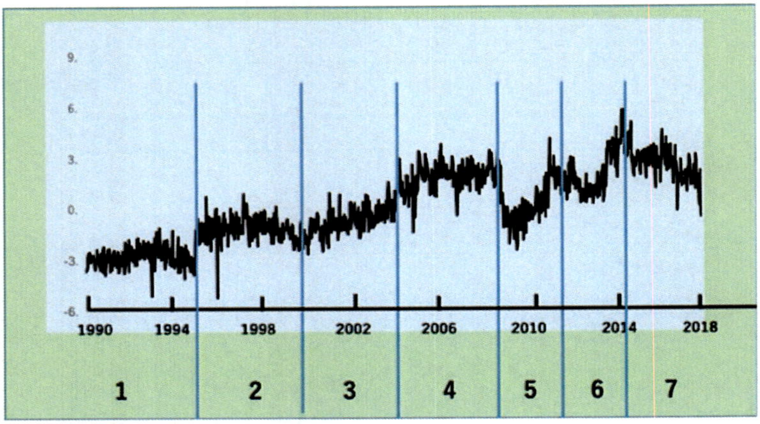

Fig. 8.5 Time series of correlated seismicity rates z_1 showing divided periods from 1 to 7 stages to investigate the successive change of their local correlation intensity of these periods

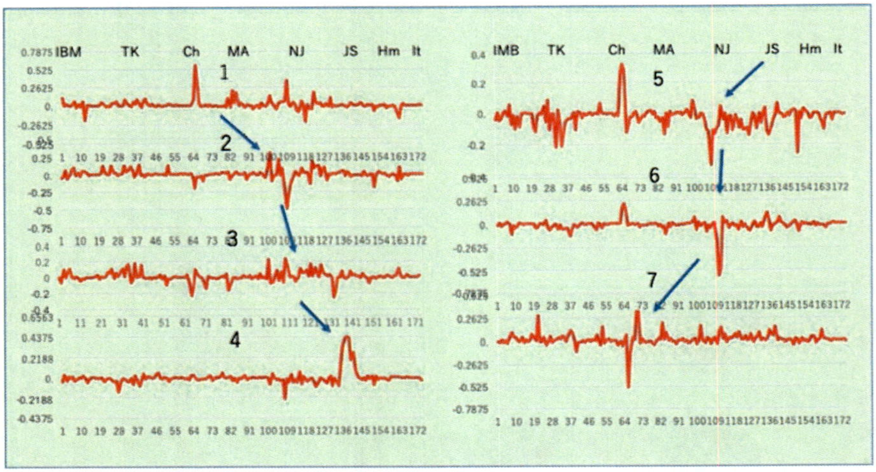

Fig. 8.6 Spectrogram of the correlation intensity of local cells (1 to 174; from IMB to MAR) from stage 1 to stage 7 in Fig. 8.5. IBM; Izu–Bonin–Mariana, TK; Tonga–Kermadic, CH; Chile, MA; Middle America, NJ; Northeast Japan, JS; Java-Sumatra, Hm; Himalaya, It; Italy. Vertical axis is the w_{1j}

Let us think about the attractors that are stable nodes in the dynamical system of the z_1 and z_2 under the constant z_3 parameter. Here, z_1 and z_2 represent the modes of the correlated seismicity rate network in the global subduction zones as shown in Fig. 8.3, and the rates of them are assumed to be followed by the minimal nonlinear differential equations consisting of follows;

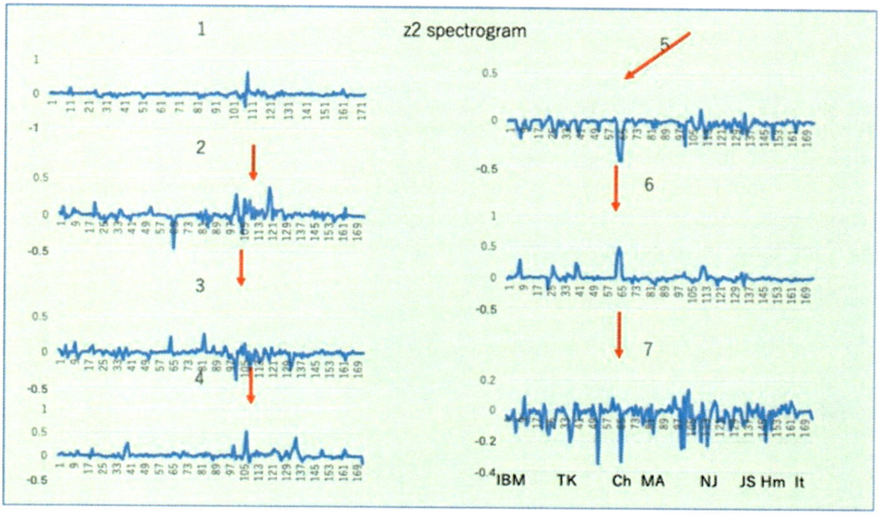

Fig. 8.7 Spectrogram of the correlation intensity of local cells (1 to 174; from IMB to MAR) from stage 1 to stage 7 in Fig. 8.5. IBM; Izu–Bonin–Mariana, TK; Tonga–Kermadic, CH; Chile, MA; Middle America, NJ; Northeast Japan, JS; Java-Sumatra, Hm; Himalaya, It; Italy. Vertical axis is the w_{2j}

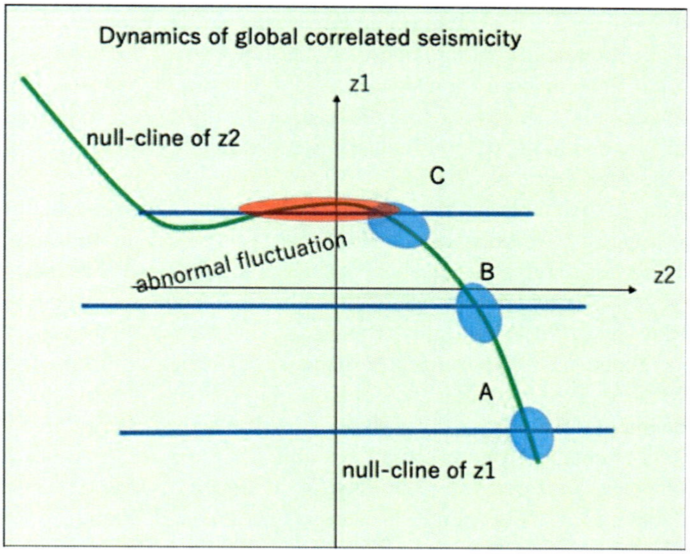

Fig. 8.8 Null-clines of minimal nonlinear dynamical equations of z_1 and z_2 in the global correlated seismicity dynamics showing the stable nodes A, B, and C by crossing two null-clines. The abnormal fluctuation takes place at the tangential relation between two null-clines

$$dz_2/dt = z_2 - a_1 z_2^3 + a_2 z_1$$
$$dz_1/dt = z_1 + b_1 \quad (8.1)$$

in which dt is the time increment in the system and a_1, a_2, and b_1 are the coefficients of the above differential equations.

Stable nodes shown in Fig. 8.3 of above ODEs can be obtained simply by the crossing points of the null-clines of above equations, these are $dz_1/dt = 0$ and $dz_2/dt = 0$, and then, the equations of null-clines hold

$$z_2 - a_1 z_2^3 + a_2 z_1 = 0$$
$$z_1 + b_1 = 0 \quad (8.2)$$

and thus,

$$z_2 - a_1 z_2^3 - a_2 b_1 = 0 \quad (8.3)$$

If we assume the simple linear ODEs as the dynamical system of the seismicity correlation z_i, there is the only single array of the stable nodes but not the bi-stable nodes implying the abnormal trends that appear in the real seismicity dynamics. The null-clines of above ODEs are shown in Fig. 8.6 in the z_1–z_2 phase space, and the large jump of the z_2 parameter is induced naturally at the crest point of z_2 third power arithmetic equation. At that point, the z_2 curve (line) should contact with the z_1 curve as the tangential relation, and then, the finite small fluctuations of z_1 or z_2 should result in the rapid transition for low z_1 to high z_1 discontinuously. The minimal nonlinear differential equation having the term of the third order of the parameter is needed and satisfied in the case of the near linear dynamics as known as the central manifold theorem (Kuramoto 2003).

As looking at the stable nodes where noises showing Gaussian fluctuation kept within, the migration of the stable nodes in the z_1 and z_2 diagram requires for the discrete change of parameters of above ODEs. Judging from that the z_1 changes progressively associated with change of strongly correlated locality cells network as shown in Figs. 8.9 and 8.10. It seems problematic that the reason why the network of strongly correlated seismicity rate cells changes drastically with time, though there are considered to be many parameters determining the above-correlated seismicity networks as internal freedom of the global seismicity dynamical system. It looks like that the mode of change from one to other stable nodes passes the route of globally murmuring phase appeared after the stage 4 of Fig. 8.6. This type of transition between stable nodes can be identified at the stages of 3 and 5 in the z_2 spectrogram (Fig. 8.7). This suggests that the transitions between two stable nodes are associated with the intermediate state that displays the global murmuring in the seismicity. Inversely, it is to say that the intermediate state is characterized by occurrence of unstable modes of global seismicity rate connection, and that the change from the stable node to this intermediate state is just the breakout of stable mode and that

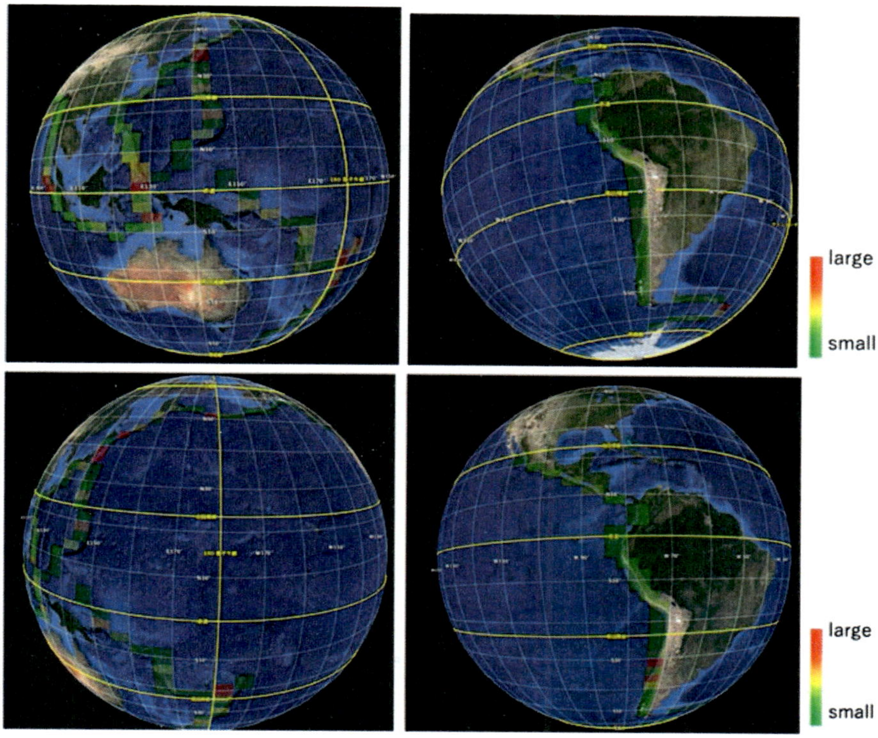

Fig. 8.9 Map showing the correlation intensity of the locality cells of the global correlated seismicity rates z_1 (top) and z_2 (bottom) on the Google Earth

from the intermediate state to next stable node is to be the self-organization of stable mode (Fig. 8.8).

So, we must forward to make a minimal dynamic equation as stated previously. The minimal dynamic equations of z_1 and z_2 should be followed by the array of the stable nodes (A-B-C) and abnormal fluctuation from the node C. The abnormal fluctuation from C to decrease of z_2 contains large variation of variances of z_2 together with z_1, suggesting that such variation does not mean the trend but the abnormally large variance of z_2 together with z_1 rather than the variances at the stable nodes of A to C.

The minimal model interpreting the above characteristic features of stable nodes and fluctuations may be expressed by the successive change of null-clines of z_1 and z_2 as shown in Fig. 8.8. In this diagram, the stable nodes A and B are to be high angle crossing between these null-clines because of small variances of z_1 and z_2, though the stable node C should be near the contact with tangential relation of two null-clines because of wide variation (large variance) of z_2 at the abnormal fluctuation trend in Fig. 8.3. Changing the z_1 null-cline with advancing time is responsible for the change of geometry in the correlated seismicity networks as mentioned above, and then, it

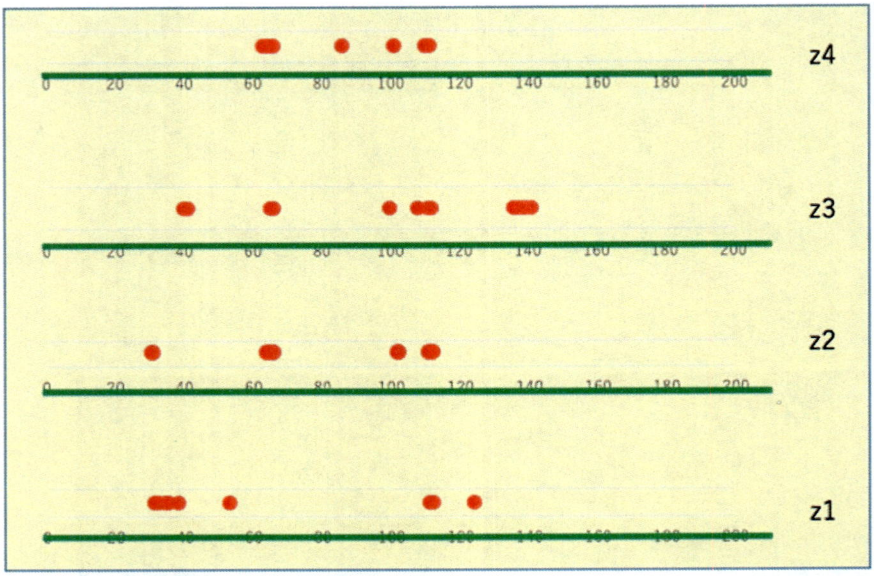

Fig. 8.10 Illustrative diagram of the representative correlated cells of the correlated seismicity rates z_1 to z_4 in the global subduction zone. Red dots show the representative correlated locality cells indicating the correlation network

leads that the coefficient such as a_1, a_2, and b_1 must be dependent upon the geometry of the network, that is the time advance.

In the discussion above, the correlated seismicity z_3 is assumed to keep constant, but the three components mapping of z_1 to z_3 (Fig. 8.3) indicates that the trend of stable nodes is surely available for the above assumption and that there is another abnormal fluctuation trend extending z_3 direction from the stable node C during the period from 2003 to 2009. The changing patterns from stable node C to the abnormal fluctuations are very similar to each other, and thus, it seems available that the same minimal dynamic equation model as that of the z_1 and z_2 space can be applied for the z_1 and z_3 space.

The data assimilation for the parameters a_1 and b_1 is carried out by the coordinates of stable nodes available for the parameters sets, and stable node C is approximated to be formed in the condition of tangential contact between two null-clines as shown in Fig. 8.8. The parameters of Eq. (8.3) are as follows;

$$
\begin{aligned}
&A \text{ node} : z_1 = -3, z_2 = -0.5 \\
&B \text{ node} : z_l = -1, z_2 = 0.7 \\
&C \text{ node} : z_1 = 2, z_2 = 0.5
\end{aligned}
\tag{8.4}
$$

Thus, the parameters in Eq. (8.3) are obtained to be as,

$$a_1 = 2.6 \pm 1.4$$
$$a_2 = -0.1 \pm 0.1$$

and

$$b_1 = 3, \quad \text{for } A$$
$$b_1 = 1, \quad \text{for } B$$
$$b_1 = -2, \quad \text{for } C.$$

The stable nodes from A to C correspond to the periods of 1990–2001, 2002–2008, and 2009–2018, respectively. According to the weight coefficients controlling the network geometry of the global seismicity rates discussed in the previous sections, the large number of network components can be seen in the periods of A and C for z_1 and for z_2 of C.

In this section, it is discussed that there are several network modes in the global seismicity rate dynamics which observe small earthquakes in the subduction zones and two of them are representative in the intensity of the correlated seismicity rates as shown in Fig. 8.11; the one is the western Pacific-Indian Ocean mode, the other is the eastern Pacific Ocean mode, and the Northeastern Japanese region is the main constituent of these two modes. In later section, the author will discuss the more general interpretation of the minimal dynamics of the correlated seismicity rate dynamics.

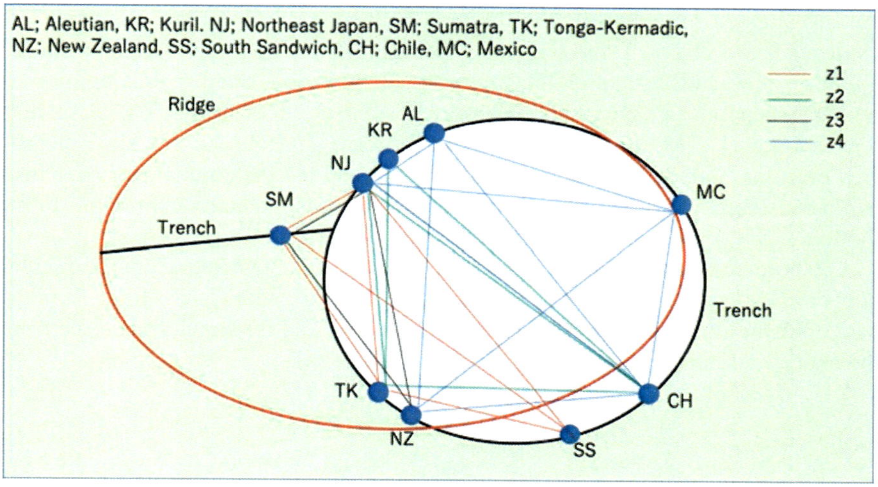

Fig. 8.11 Illustration of the correlated seismicity network in the global seismicity dynamics

8.2 Synthetic Coherency of Seismicity Dynamics by Slider Block Model

For the spatio-temporal seismic activity features of the inhomogeneous plate boundary zones, the two-dimensional array of multiple connected slider blocks models have been proposed by several authors (Brown et al. 1991; Bak and Tang 1989; Ito and Matsuzaki 1990). They considered that the cellular automaton model can be applicable for the stochastic nature of the earthquakes generation, and they succeeded in reproducing the relation of the earthquake magnitude and their frequency likely to Gutenberg and Richter law. They insisted that the power-law type relation of them as the scale-free mechanics means the physical phenomena of the self-organized criticality as proposed by Bak and Tang (1989). Their models are basically composed of the spring–blocks connected each other on two-dimensional network.

The two-dimensional network of many spring and slider blocks can be equivalent with the 2D cellular automaton because of the displacement of every cells that are controlled by the Hook's force resulted from four nearest neighboring cells as like as the state transition of the cell by the rule of 2D configuration of the nearest neighboring cells stated in the cellular automaton. Therefore, it seems to be applicable for the seismicity that there should be various invariant structures of the spatio-temporal patterns of the development of such type cellular automata divided by the controlling rules equivalent with physical interaction between the cells. In this study, the global seismicity fluctuation patterns of the whole plate boundary zones are investigated by means of the multiple slider–spring block models which are connected as a chain of one dimension. In this model, the chain is analog to the global plate boundary, and on the chain, the block is connected with hanging wall crust by shear spring and with neighboring blocks by shear spring. The block also received the viscous stress from the asthenosphere controlled by its slip velocity. However, the boundary stress between the plate and the underneath asthenosphere is not considered because of the plate velocity which is gradually continuous to the motion of the whole mantle convection.

Let us consider the model that the slider blocks are connected to form the circular network as shown in Fig. 8.12. The blocks are operated by the Hook's force from the hanging wall crust and from the neighboring blocks and by the viscous force from the underneath mantle as follows;

$$\begin{aligned} F_i &= -k_{II} x_i - k_s(x_i - x_{i+1}) + k_s(x_{i-1} - x_i) - r v_i \\ dv_i/dt &= F_i \\ dx_i/dt &= v_i \end{aligned} \quad (8.5)$$

in which k_u and k_s are elastic shear coefficients and r is the viscosity of the mantle.

In order to obtain the characteristic features of spatio-temporal patterns of the stress and strain distribution which should govern the seismicity rate patterns along on the plate boundary zones, the total number of the blocks connected with circular

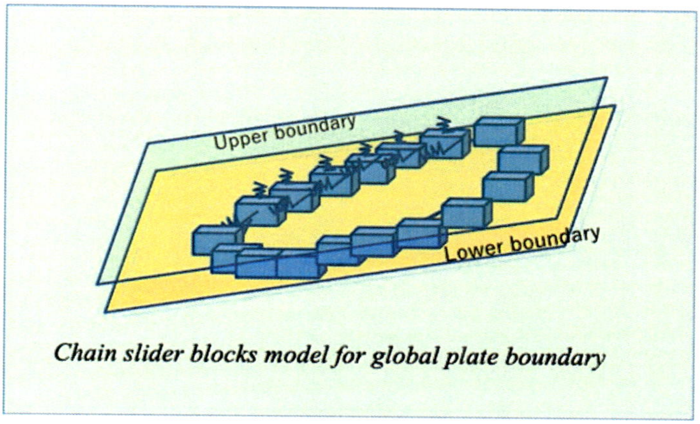

Fig. 8.12 Circular connected slider–spring block model showing the subduction zone slippages at the plate boundary. The upper boundary is of the subduction plate and overriding plate, and the lower one is with the asthenosphere underneath the subduction plate

network is the case of 100, and the initial small displacement x_i is considered to be random in the range of 10^{-6}. Furthermore, the initial velocity of the block v_i is given as zero. In addition, the effective elastic constants k_u and k_s and viscosity h should be controlled by the porosity and water content of rocks, but in this study, they are approximated to be constant for avoiding the complexity in this study.

The simulation experiments were carried out as follows: the initial displacement is given by randomized number of $-10^{-6} < x < 10^{-6}$ and the initial velocity to be zero. The displacement of x_i of each step is followed by the above equation and is added by the randomized number, dx of

$-0.00001 < dx < 0.00001$. The slider blocks lined in a circle network move step by step followed by above differential equations. The time step in this simulation is taken to be 10^{-5}, and total steps are 5000.

The shear modulus of k_u and k_s are assumed 4.5 and 4.0 in the non-dimensional form, respectively, and the viscosity r is 0.01 in this study (Table 8.1). The experimental results of the time series of displacements of various blocks and their data cloud in the diagram of three representative blocks are shown in Figs. 8.13. The times series of them are processed, and then, the correlated displacements and correlated stresses of the blocks are obtained by means of the probabilistic PCA algorithm

Table 8.1 Experimental parameters of slider-block simulation: spring constant; K1, K2, viscosity; r, initial displacement; *d*, initial velocity; *v*, time step; dt, total step; t

	K1	K2	r	$d \times 10^{-6}$	v	dt	Total t
Experiment-1	4.5	4.1	0.01	$-1 \sim 1$	0	10^{-5}	5000
Experiment-2	4.05–4.55 random	4.0	0.01	$-1 \sim 1$	0	10^{-4}	5000

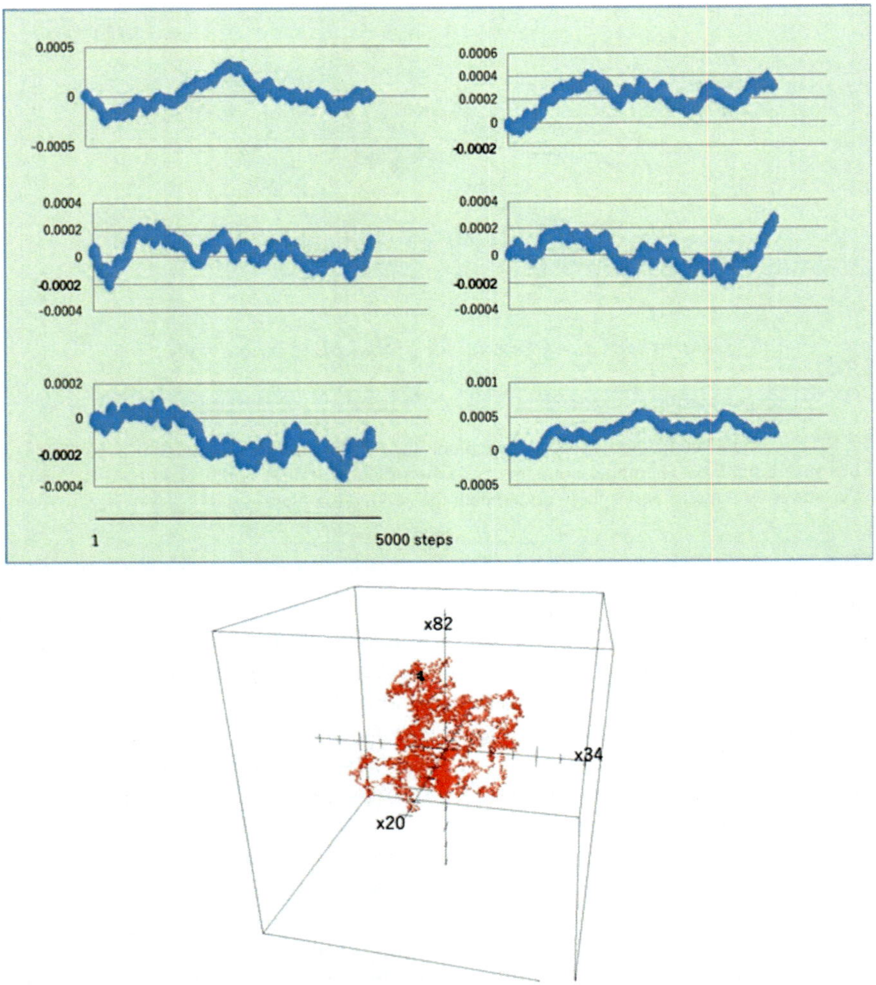

Fig. 8.13 Representative time series of the displacement of representative blocks (simulation; 5000 steps) on the circular connected slider–block model. The vertical axis is the normalized displacement, and horizontal axis is the time step. Bottom; data cloud in the diagram of displacement of cells x20, x34, and x82 in the long-term simulation of the slider block model

described in the previous sections. The time series of the correlated displacement and stress are shown in Figs. 8.14 and 8.15.

The geometry of the network is essential for the characteristic features of the correlated patterns of the stress and the displacement of the blocks as shown in Fig. 8.16,

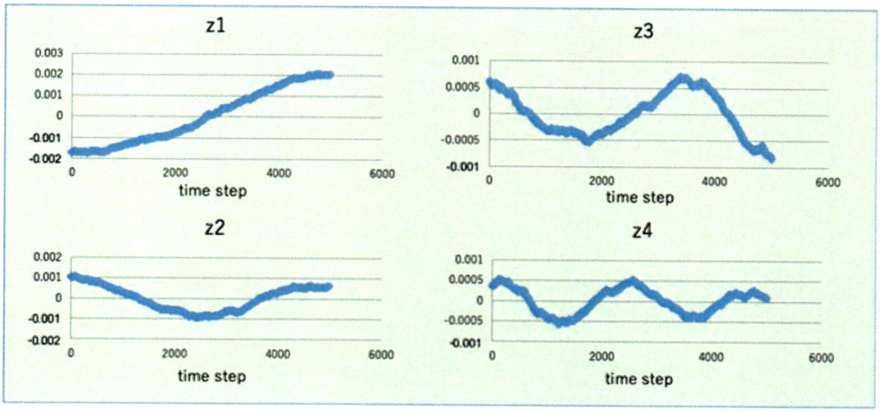

Fig. 8.14 Time series of the correlated displacement of simulated blocks z_1 to z_4 of the circular connected blocks

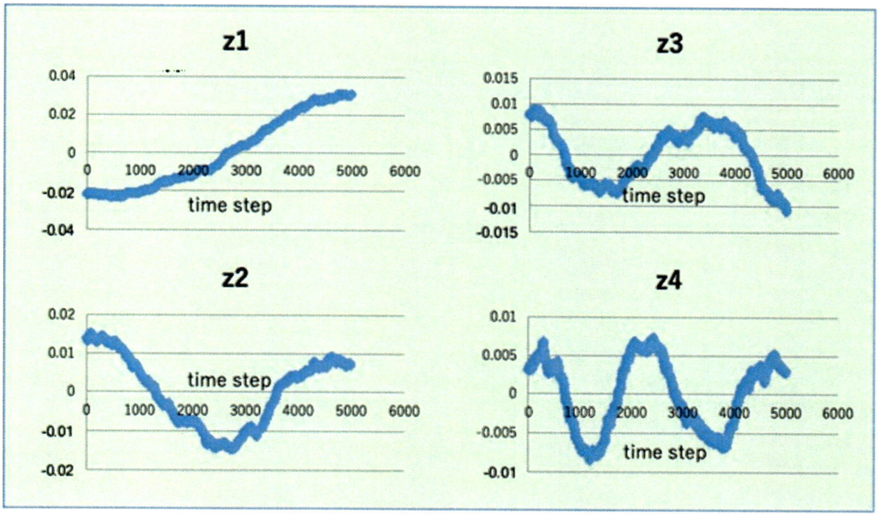

Fig. 8.15 Time series of the correlated force (stress) of the slider blocks z_1 to z_4 in the simulated system

judging from comparison of the natural correlated seismicity rate patterns discussed in this book. The results of the correlated stress distribution patterns are shown in Figs. 8.14 and 8.15, suggesting that there are several components of cyclic fluctuations but every correlated stress fluctuation associates with the distinctive phase shift. In addition, the weight coefficients of the correlated stress components (Fig. 8.17) are often semi-periodic, but the neighboring cells have the different sign, showing the negative and positive stress active for the neighboring cells. However, the differential

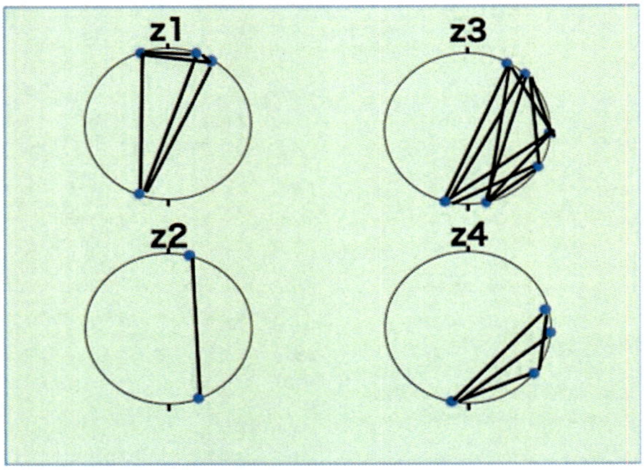

Fig. 8.16 Illustration of network patterns of the correlated displacement, z_1 to z_4 of the circular connected slider blocks in the simulation experiments

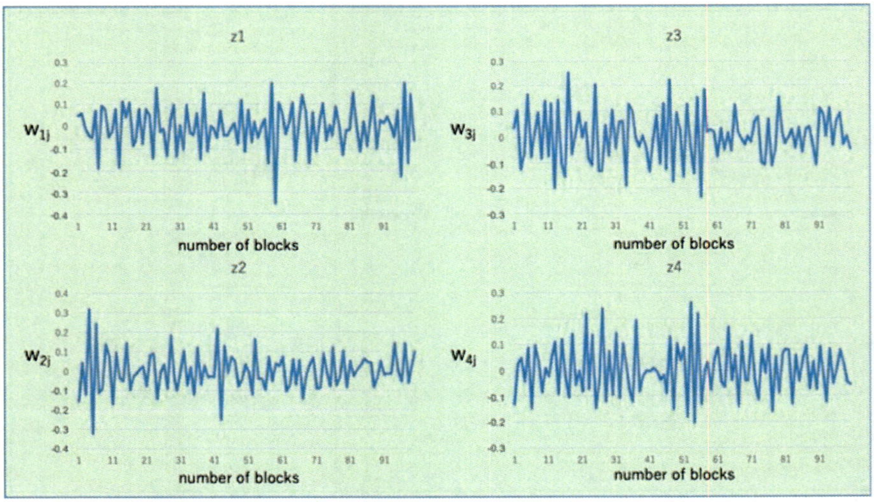

Fig. 8.17 Linear diagram of the correlation intensity (w_{ij}) of circular connected slider blocks, z_1 to z_4 in the simulation

stress intensity of their neighboring cells affects the similar behavior of crack generation and propagation. It may conclude that the circular chained spring and slider block model may give the essential patterns of the correlated displacement and stress even in the initial conditions of random displacement and zero velocity blocks. The time scales of the periodicity of the correlated stress seem to be determined by the scales of the block size and the displacement length per unit step. In this simulation study, the total step number reaches 5000, and this seems to be equivalent to the time scale about 500 years, judging from that the accumulation of the displacements of many blocks is roughly 2000 steps, and the natural plate boundary zones have repeated cycle about several hundred years of displacement relative to the hanging wall crust.

The phase maps of the correlated stress intensity in the chained slider blocks are characterized by the mode separations of z_1 to z_4 components as shown in Fig. 8.18, suggesting that the phase map of the z_1 and z_2 means the trajectory of low to high z_1

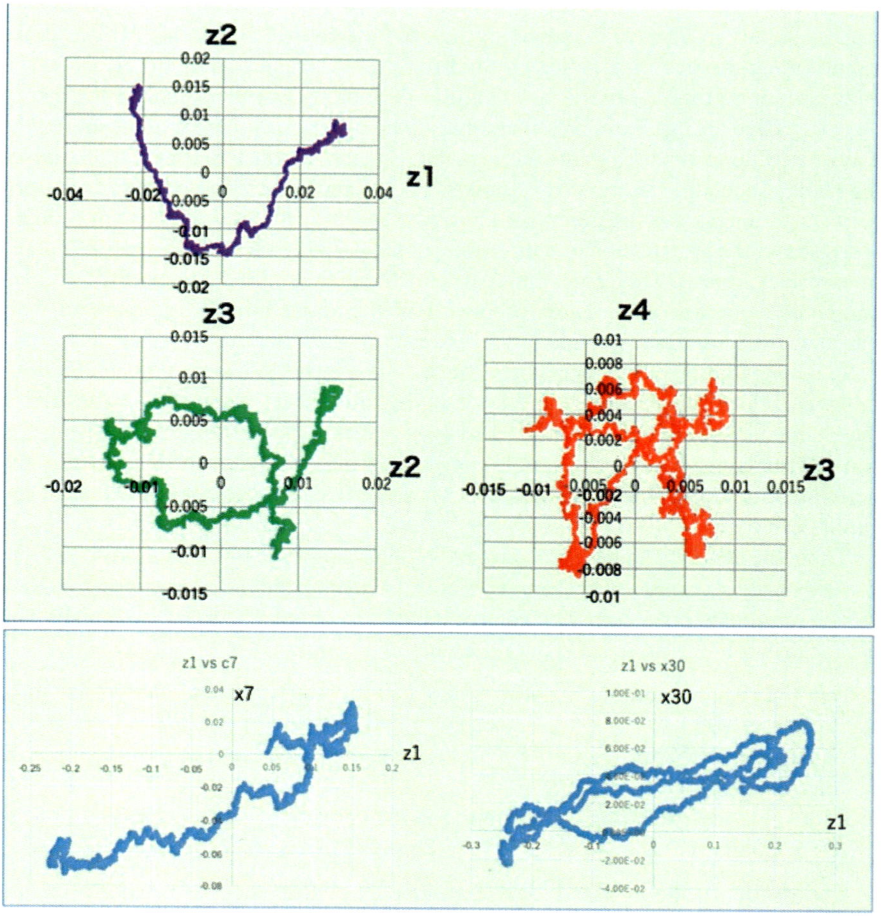

Fig. 8.18 Relations of the correlated displacement of connected slider blocks, z_1, z_2, z_3, and z_4 showing the coherent time series of their cyclic variations. Bottom; simulated relations of correlated stress z_1 and the cell stress xi in the circular connected slider block model. Left; short simulation, right; long simulation

with cyclic variation of z_2, but in the z_2 and z_3, and also in the z_3 and z_4 phase map, the simulated points look like the Lissajous figures of them. Therefore, it seems that the correlated stress and displacement time series are the periodic signals with phase shift and different frequency. Furthermore, the relationships between the correlated force and individual force on the blocks are shown in Fig. 8.18, suggesting the possibility of inversion from the correlated force to the individual slider block force as discussed in the preceding chapter using the main sequence of the correlated seismicity and individual seismicity.

The correlated spring–blocks sets can be seen in the weight coefficients intensity map as shown in Fig. 8.16, indicating that the different associations of the blocks behave the periodical fluctuations with different phase and periodicity. However, it is important that the z_2 and z_3 have similar periodicity but their phase shift means the successive change from one to another in the time series. It should mean that the wavy migration from the z_2 time series to the z_3 one is responsible for the transition of the block association from z_2 to z_3 modes. There should be observed in the relations between z_1 and z_2. On the other hand, the z_3 mode is different with those of z_1 to z_2, and it shows the localization of the z_2 mode of the periodic fluctuation appeared in the time series of the z_2. Therefore, it may be concluded that the correlated displacement and force represent the coherent behavior of the slider blocks motions and their internal stress conditions which probably result in the active seismicity rates.

In the simulation experiments mentioned above, the obvious periodic fluctuations in the correlated displacement and force in the time series seem to be derived from the elementary mechanics of the studied system that is the network of the spring and slider units controlled by the simple oscillator. Thus, it is easy to imagine that the periodic structure of the correlated motion and force is responsible for that of the simple oscillators.

Then, the synthetic data of the network system composed of the random variation, the correlated sinusoidal wave, and non-correlated sinusoidal time series in the 20 × 20 cell matrix are processed by means of the PCA method discussed previously. The data matrixes are simulated by the following algorithm:

1. put the random number between 0 and 1 in 20 × 20 cells and every time steps from 1 to 1000,
2, sinusoidal synchronous variation in the cells along.the tenth row of the 20 × 20 matrix,
3, sinusoidal non-synchronous variation in the cells along the tenth column of the 20 × 20 matrix.

The results of the synthetic data experiments of the correlated pattern test are shown in Figs. 8.19, 8.20 and 8.21. It is obvious that the synchronous and non-synchronous variations of the cells embedded in the random matrix as shown in Fig. 8.19 can be clearly transformed to be sharp characteristic features given initially, but the characteristic features are noisy even. Thus, it concludes that the deterministic and sparse PCA decomposition methods are available for investigation of the characteristic features of the multiple parameters time series analyses as discussed in the previous chapters.

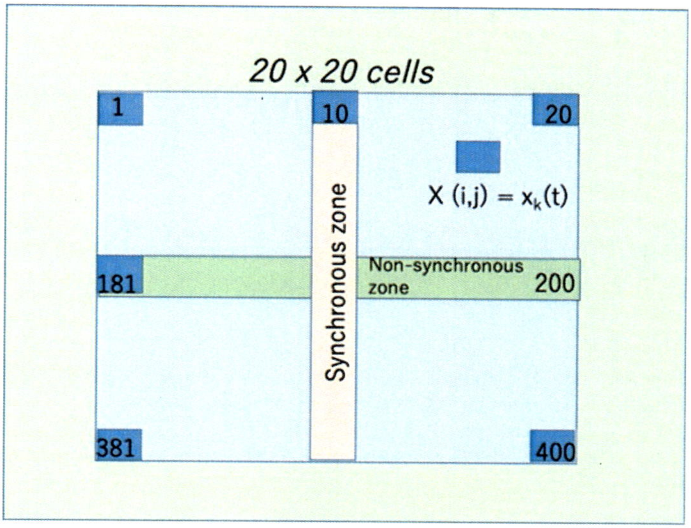

Fig. 8.19 Illustration of the simple synthetic data set of the synchronous, non-synchronous, and random signals embedded in the systematic cell configuration

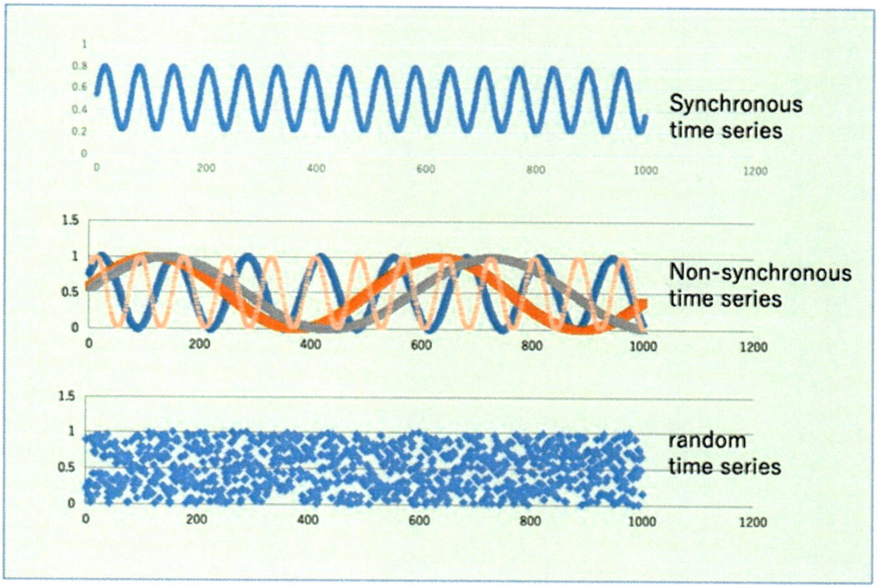

Fig. 8.20 Synthetic signals of the simple configuration of the synchronous, non-synchronous, and random time series

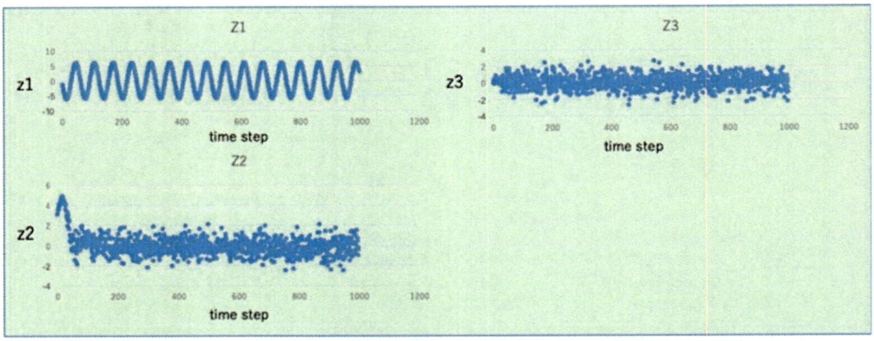

Fig. 8.21 Time series of the correlated signals z_1 to z_3 in the simple synthetic data set shown in Fig. 8.19, showing the synchronous pattern of z_1 and random one of z_2 and z_3

References

Bak P, Tang C (1989) Earthquakes as a self-organized critical phenomenon. J Geophys Res 94:15635–15637

Brown SR, Scholz CH, Rundle JB (1991) A simplified spring-block model of earthquakes. Geophys Res Lett 18:215–218

Ito K, Matsuzaki M (1990) Earthquakes as self-organized critical phenomena. J Geophys Res 95:6853–6860

Kuramoto Y (2003) Chemical Oscillations, Waves, and Turbulence, Dover, 340 p

Chapter 9
Seismicity Dynamics Model of Global Earth and Japanese Islands Region

Abstract The minimal dynamics of the correlated seismicity rates in the global and regional systems is classified into the linear case and the nonlinear one: The linear case of the correlated seismicity dynamics is simply expressed by the time derivatives (rate) of correlated seismicity with linear combination of several components of the correlated seismicity rates. On the other hand, the nonlinear case is rather complicated as the minimal nonlinear differential equations such as bi-state bifurcation model.

Keywords Minimal model of global and japanese seismicity dynamics · Null clones of the seismicity dynamic equations · Annual variation of partial b-value change

9.1 Minimal Model of the Global and Japanese Seismicity Dynamics

As stated in the previous chapters, the data matrix of the global seismicity rates is possibly transformed into the correlated seismicity rates z_i by means of the deterministic and stochastic principal component analyses. Especially, the stochastic PCA with LASSO is strong tool of the dimension reduction method using sparse modeling (SpM) in order to obtain the characteristic features of the high dimension network system. In high dimension dynamical system, the uniquely determined low dimension characteristic parameters are thought to be expressed by following multiple ordinary differential equations in the simple way: The z_i of the newly obtained variable set hold;

$$dz_i/dt = F(z_j) + N_i$$
$$dU/dz = F \quad (9.1)$$

Electronic supplementary material The online version of this chapter (https://doi.org/10.1007/978-981-15-5109-3_9) contains supplementary material, which is available to authorized users.

M. Toriumi, *Global Seismicity Dynamics and Data-Driven Science*, Advances in Geological Science,
https://doi.org/10.1007/978-981-15-5109-3_9

in which U is the potential function of z and N_i is the Gaussian noise, respectively. This type of equation is called as Langevin equation of the stochastic differential equation.

The potential function of U is commonly expressed by the polynomial even function of z_i as follows;

$$U(z_j) = \Sigma(a_{ij} + b_{ij}\, z_i^2 + c_{ij}\, z_i^4 + \ldots .) \tag{9.2}$$

where a_i, b_i and c_i are coefficients of the polynomial expansion of U. In this equation, summation is made by i. If F is the case of the linear function, the simple dynamics holds as

$$dz_i/dt = 2\,\Sigma(b_{ij}\, z_j) + N_i \tag{9.3}$$

or

$$dZ/dt = A\,Z + N$$

where Z is the vector of $\{z_i\}$ and A is the coefficient matrix of $\{a_{ij}\}$. Considering that above multiple ordinary differential equations are not degenerated, the null-clines of above equations give the stable node of solution as follows;

$$AZ + N = 0 \tag{9.4}$$

If considering the change of the mean value of the Gaussian noise, N, the single stable node of the above ODF should change with time, but the stable node keeps the single value.

On the other hand, if considering the double stable nodes derived from bifurcation of the equation above, the nonlinear differential equation is needed for modeling the dynamics concerning the correlated parameter, z. In other words, because of the z meaning the strongly correlated mode of cells network about seismicity rate, the double state nodes of z parameters imply the transition from one mode to another mode of the correlation network type. The transition between the strongly correlated network types is already found in the simulation experiments of the spring–slider blocks network proposed previously. In those simulations, it concluded that the four modes of the block network change their intensity and their time series change periodically with different phase. Therefore, the network mode changes itinerantly from one to another, and it is possibly said that the dynamics of the correlated network mode is governed by the correlated parameters, z, in the global and regional seismicity.

Let us refer the results of the correlated patterns of the global seismicity rates discussed in the previous sections. As seen in Fig. 9.1 and also in Fig. 5.19, the migrated averages of the correlated seismicity rates, z, of the global seismicity along the subduction zones in the world reveal the time series of step like change, semi-periodical change and random noise of Gaussian type. Furthermore, the network features of the strongly correlated locality cells are also shown in the network diagram

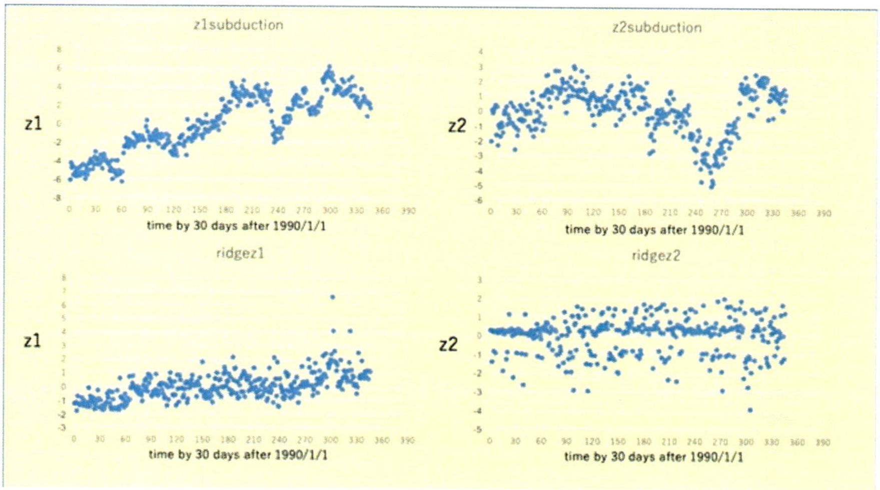

Fig. 9.1 Time series of the correlated seismicity rates z_1 and z_2 of the global subduction zone and oceanic ridge zone

modified from geographical map of the earth (Figs. 5.27 and 5.28). In this seismic correlation network diagram, it is obvious that the z_1 and z_3 networks are localized in the western Pacific and Indian Ocean segments, but the z_2 and z_4 are extended in the whole Pacific Ocean region.

The network geometry of z_1 to z_4 gradually changes from 1990 to 2018 as seen in Fig. 5.27, and it concludes that during the period of 1990–2004, the z_1 is dominant pattern though during the period of 2004–2018 the z_1 and z_2 become alternatively dominant. The network patterns of z_3 and z_4 seem to change in the mode likely to the anti-phase. Therefore, it may be concluded that the western Pacific and Indian mode of high seismic activity state of the network appear alternately to the whole Pacific mode of that network. The key stone of the alternated activity of these networks is the locality cells of Northeast Japan (NEJ) judging from number of the correlation network connections among cells. The transition in the correlation network from the z_1 to z_2 under the fluctuated z_3 and z_4 is most important phenomenon in the global seismicity dynamics. In the next sections, the physical process of this phase transition in the correlation network of the global seismicity dynamics will be discussed using the attractor diagram of (z_1, z_2, z_3) space.

On the other hand, the spring–slider blocks network model studied previously is the key to understand this type of geometry transition of the network system. As seen in Fig. 8.16, the z_1 type network appears in the initial and the final stages of the simulation together with the z_2 type network but disappears in the middle stage in which z_2, z_3, and z_4 networks appear. These modes of network display the cyclic change of their intensity with synchronous style and anti-synchronous one. Considering that the spring–slider block circular network is governed by the simple spring and viscous force between blocks and hanging wall and underneath ground to behave as the simple oscillator, the transition from the one correlation network to another network is governed by multiple interaction of many oscillators. Therefore, the behavior of the network of the correlated oscillator may be expressed by the similar oscillators dynamics with various interactions between the modes of their network. Thus, the intensity of the correlated network of oscillators may be formulated as follows if we consider the z_1 to z_3 network modes;

$$\begin{aligned} z_1 &= a_1 \cos(k_1 t + p_1) + \text{noise} \\ z_2 &= a_2 \cos(k_2 t + p_2) + \text{noise} \\ z_3 &= a_3 \cos(k_3 t + p_3) + \text{noise} \end{aligned} \tag{9.5}$$

in which k is the frequency, and p is the phase shift, and a is the amplitude. This implies that each mode of the network slider blocks corresponds independently to an oscillation with the ordinary frequency and phase shift. As a result, the attractors in the (z_1, z_2, z_3) diagram are likely to the Lissajous figures of independent cyclic signals as mentioned previously. Besides, the higher network components of z_5 appear obviously the Gaussian random noise as stated in the previous sections. It leads, therefore, that the general large-scale network dynamics of the multiple connected slider blocks with various noises can be formulated by the linear combinations of the independent oscillators having parameters of their frequency, phase shift, and amplitude. Furthermore, it is very important that the inversion analyses from correlated parameters of z to the stochastic time series of original nodes of every blocks are possibly carried out with various Gaussian noises. In the following sections, the simple inversion problems of the global and regional seismicity dynamics will be discussed explicitly.

In the case of the global network of correlated seismicity rates, the correlation network modes mentioned above are mapped in the (z_1, z_2, z_3) diagram with successive time from 1990 to 2018. It is suggested that the z_1 trend displays a gradual increase with various fluctuations, and the z_2, on the other hand, shows a weak pattern of the periodicity. Thus, the attractor in the z_1 and z_2 diagram is very likely to the simple pattern of the Lissajous figures, which show the basic pattern presented previously in the case of network slider blocks as shown in Fig. 8.18. However, the Lissajous figures of the globally correlated seismicity rates z_1 to z_4 display some magnitude of large intensity of noises, and thus, they look like different modes of coupling between every correlated mode. But, it seems obvious that every modes of correlated seismicity rate network show periodic to subperiodic change in their time series, and thus, it is natural that the diagrams of z_1–z_2, z_2–z_3, and z_3–z_4 appear likely the Lissajous figures with random noises as mentioned above.

The reason why dominant modes of globally correlated seismicity network indicate the periodic features of the time series likely to the case of the behavior of many spring–slider blocks network model simulated earlier may be derived from the global seismicity dynamics being composed of the elastic, plastic, and slippage elements. It may lead in that case that such composite many body systems may be controlled apparently by many oscillators as seen in simulated slider blocks system. In addition, it is very important that the correlated network parameters of z are simply expressed by linear combination of those of the original locality cells. Then, for the parameters z are mutually independent and have the unique frequency and phase shift, the model system of them are reconstructed by the following equations,

$$\begin{aligned} dz_1/dt &= f(z_1, z_2, z_3) + \text{noise1} \\ dz_2/dt &= g(z_1, z_2, z_3) + \text{noise2} \end{aligned} \tag{9.6}$$

in which f and g mean the polynomial equations of z_1 to z_3, and noise1 and noise2 are the Gaussian noise with mean $= 0$ and variance σ^2. Therefore, the dynamical system of the globally correlated seismicity is possibly determined by the assimilation of above equations in the diagrams of z_1–z_2, z_2–z_3, and z_3–z_4 as shown in Fig. 9.2. The assimilation should be carried out from the mean values of available clusters obtained from EM cluster program (mclust) on R mentioned previously. Actually, the z_4 and higher z components seem to be noisy, and thus, they are included in the Gaussian noise term in above equation. Therefore, the minimal dynamics should be determined by the polynomials of z_1–z_3, and the coefficients of their terms should be statistically determined by three sets of them for A, B, and C in Fig. 9.2. For different segments of S_1 and S_2 which are abnormal trajectories at 2004 Sumatra and 2011 Tohoku-Oki giant earthquakes, these are excluded in the studied dynamical equation. But it should be noted that these abnormal trajectories are started at the C clusters in Fig. 9.2, suggesting that it should be a special cluster for the dynamics. Considering the change from A to C through B with advancing the time as shown in Fig. 9.3, these clusters should be stable.

The clusters of A, B, and C are satisfied with stable nodes of dynamical equations above, for the stable nodes that are determined by the null-cline equations of above ODEs. Now let us assume the two parameters' dynamics as follows;

$$\begin{aligned} dz_1/dt &= f(z_1) - g(z_2) + \text{noise1} \\ dz_2/dt &= g(z_2) - h(t) + \text{noise2} \end{aligned} \tag{9.7}$$

in which g is the function of z_2.

Considering the null-cline of above equation, the stable node is determined by the following equation,

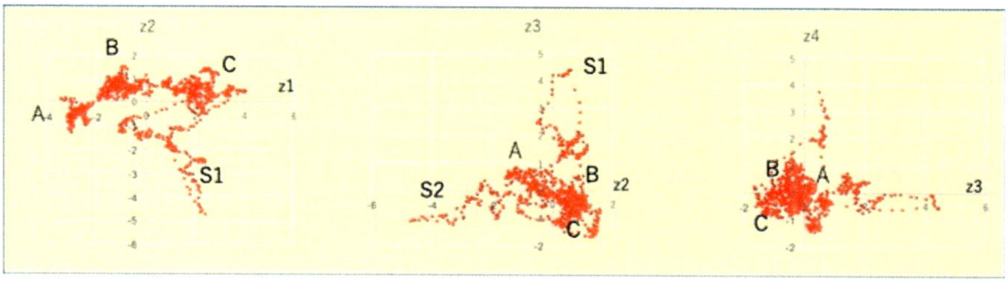

Fig. 9.2 Diagrams of the correlated seismicity rates z_1–z_4 in the global subduction zone. Stable clusters A, B, and C associated with scattered clusters S_1 (2004) and S_2 (2011)

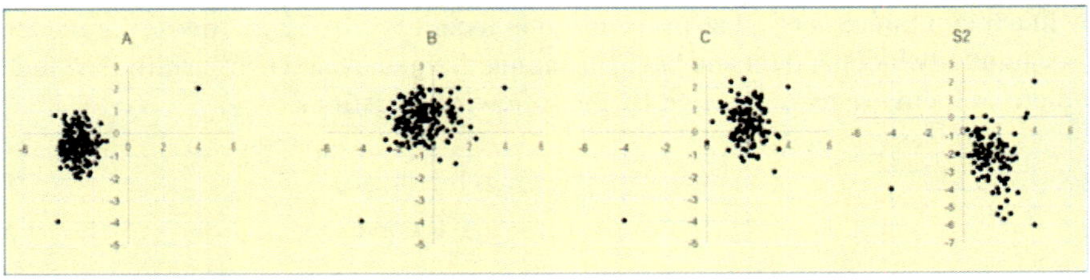

Fig. 9.3 Diagrams of the correlated seismicity rates z_1 and z_2 of the global subduction zone in the successive change from A to C and scattered S_2 with increasing time

$$\begin{aligned} f(z_1) - g(z_2) &= 0 \\ g(z_2) - h(t) &= 0 \end{aligned} \tag{9.8}$$

By the way, there seems the transition between A and B, and B and C nodes as shown in Fig. 9.3. Thus, it should be available that the $f(z_1)$ may be expressed by the minimal nonlinear equation judging from the center-manifold reduction method (Kuramoto 2003) as follows;

$$f(z_1) = c\, z_1^3 - d\, z_1$$

and

$$g(z_2) = z_2 \tag{9.9}$$

For this type of dynamics of z_1, it leads that there are stable nodes and unstable node (separatrix) in the above ODE system. The representative features of these nodes in the diagram of z_1 and z_2 are shown in Fig. 9.4. In this diagram, the transition from the nodes of B to C is to be discontinuous, and it must occur at the tangential condition of null-clines mentioned above. Here, the function of $h(t)$ should be periodic behavior because of periodic z_2 and z_3 as noted in the previous sections, and so the changing trend of stable nodes from A to C through B should be followed by D according to decrease of $h(t)$, or by E with increase of $h(t)$. Therefore, the parameters of above ODE are possibly determined by data assimilation of the stable nodes defined above. But, it may need that the $h(t)$ should be estimated as the polynomial function of the z_2 and z_3 with randomly changing z_4. This problem will be discussed in the next chapter.

Remember the transition from B to C meaning the change of the global correlated seismicity network geometry from the Pacific type to Indian-Pacific type. The former type is represented by z_2 mode and the latter by z_1 mode as seen in Fig. 8.11. The minimal dynamic model requires for the two states conditions of z_1 mode dominant and z_2 mode dominant stages. Furthermore, the fluctuation at the nodes A and B are rather small, but at the node C, it increases abruptly as shown in Fig. 9.3.

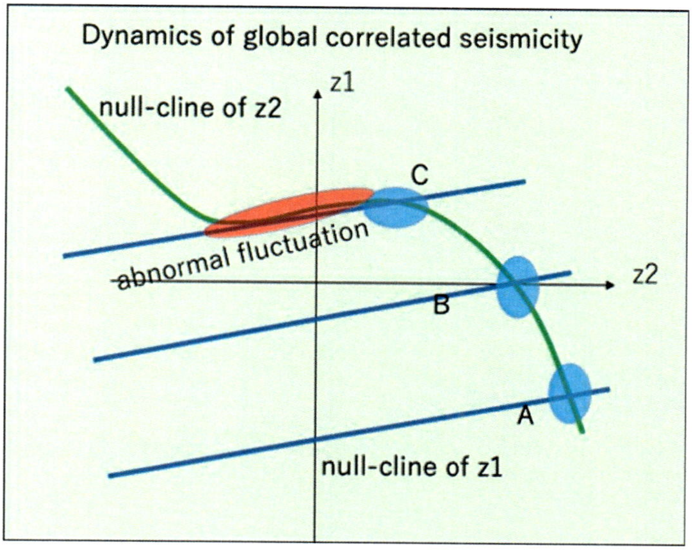

Fig. 9.4 Null-clines of minimal nonlinear dynamical equations of z_1 and z_2 in the global correlated seismicity dynamics showing the stable nodes A, B, and C by crossing two null-clines. The abnormal fluctuation showing the scattered cluster in Fig. 9.3 takes place at the tangential relation between two null-clines

9.2 Minimal Dynamic Model of Japanese Correlated Seismicity

In the downscaling dynamics of Japanese islands region, the similar situation with the globally correlated seismicity dynamics should be satisfied in the behavior of z_1–z_4. First, dynamical equations of z_1 and z_2 with noisy z_4 and the higher are written likely to the global case as follows,

$$\begin{aligned} dz_1/dt &= c\,z_1^3 - d\,z_1 - z_2 + \text{noise1} \\ dz_2/dt &= z_2 - h(t) + \text{noise2} \end{aligned} \tag{9.10}$$

in which f should be nonlinear type as like as the case of the global correlated seismicity dynamics. The function of g needs basically polynomial function of z_2 and z_3 that are dependent on the time, and thus, it is to be a periodical function of the time. Considering that there are four stable nodes and transient state in the diagrams of z_1–z_2 and the sequential tracks from node A to C and D are present (Fig. 9.5), the stable nodes of the above minimal nonlinear equation should be obtained as follows likely to the global case as shown in Fig. 9.6;

$$c\,z_1^3 - d\,z_1 = 0$$

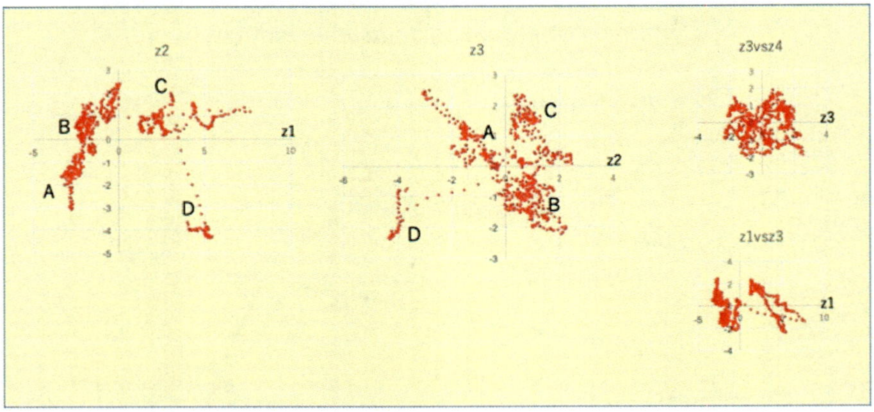

Fig. 9.5 Diagrams of the correlated seismicity rates z_1 to z_4 in the Japanese islands region showing the successive transition from A to B, B to C with abnormal scattered cluster with 2011 Tohoku-Oki earthquake, and C to D

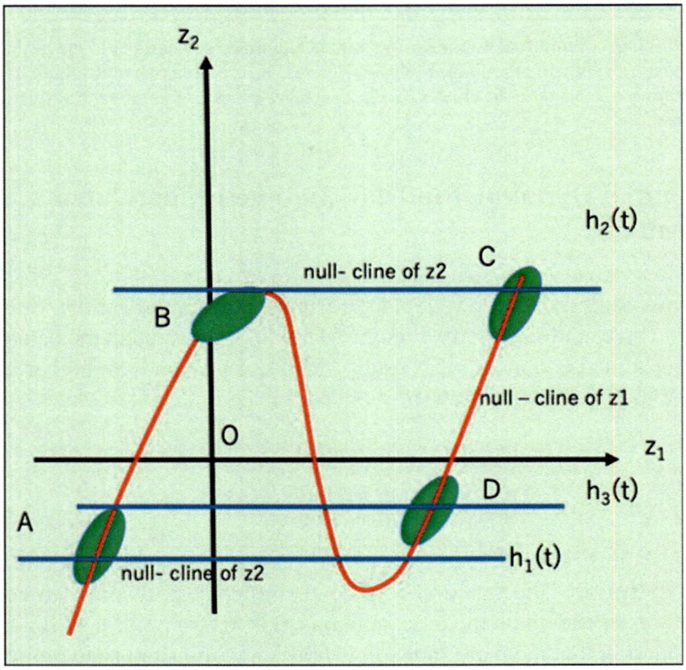

Fig. 9.6 Diagram of the null-clines of the correlated seismicity rate dynamics of z_1 and z_2 in the Japanese islands region, showing the stable nodes as A, B, C, and D. At node B, the both null-clines are tangential with each other

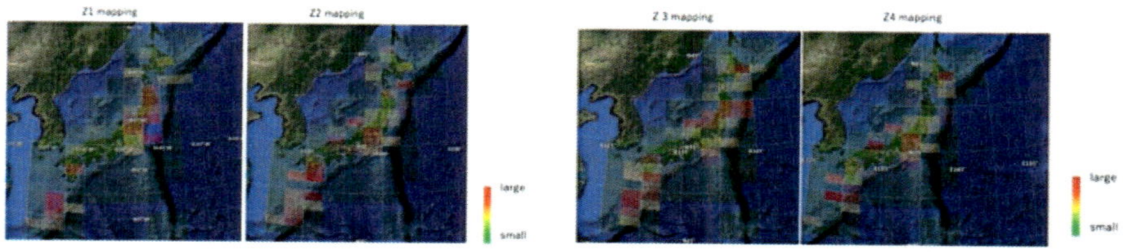

Fig. 9.7 Right, map showing the highly correlated locality cells of the correlated seismicity rates z_3 and z_4. Left, map showing the highly correlated locality cells of the correlated seismicity rates z_1 and z_2

$$z_2 - h(t) = 0 \tag{9.11}$$

Furthermore, $g(t)$ seems to be periodical against the time, for z_2 varies largely against that of z_1 change at the transition gap from A to B. It is to be noted that the transition from B to C is associated with rapid jumping variation showing the overshoot to the higher region of z_1. This represents the Tohoku-Oki giant earthquake of 2011/3/11 that means probably the transient state near C node.

In this correlated seismicity dynamics of Japanese islands region, the correlation of the seismicity networks changes from z_1 mode to the z_2 one according to the intensity of these components. It seems that the z_1 mode represents the network of NE Japan and Okinawa-S. Kyushu and the z_2 does that of S. Kyushu, Nankai, Tottori, Kanto, NE Japan, and Hokkaido as shown in Fig. 9.7. In other words, the gradual change of the z_2 mode is responsible for the rapid change of the intensity of z_1 mode which means probably the occurrence of a phase lock of the cyclic deformation of the Pacific and Philippine Sea plates.

The transition from B to C should be associated with large fluctuation because the stable node ultimately closes to the unstable node at the tangential relation between $h(t)$ and $f(z_1)$: The fluctuation of the z_1 signals becomes growing relative to the z_2 fluctuation so as to be nearly zero of gradient $f(z_1)$ against z_1. As discussed in the earlier sections, the short-term fluctuations of z_1 and z_2 subtracting the secular trends of them display obviously the subannual to tri-annual periodic variations. Therefore, the successive change of such fluctuations of z_1 and z_2 is to be compared before and after the transition from B to C. It seems clear that there is the wide variation of the z_1 short-term fluctuation against the z_2 one at the stage of A, compared with those at the other stages: The starting of the wide z_1 fluctuation is about 2010 to 2010 of C as shown in Fig. 9.8.

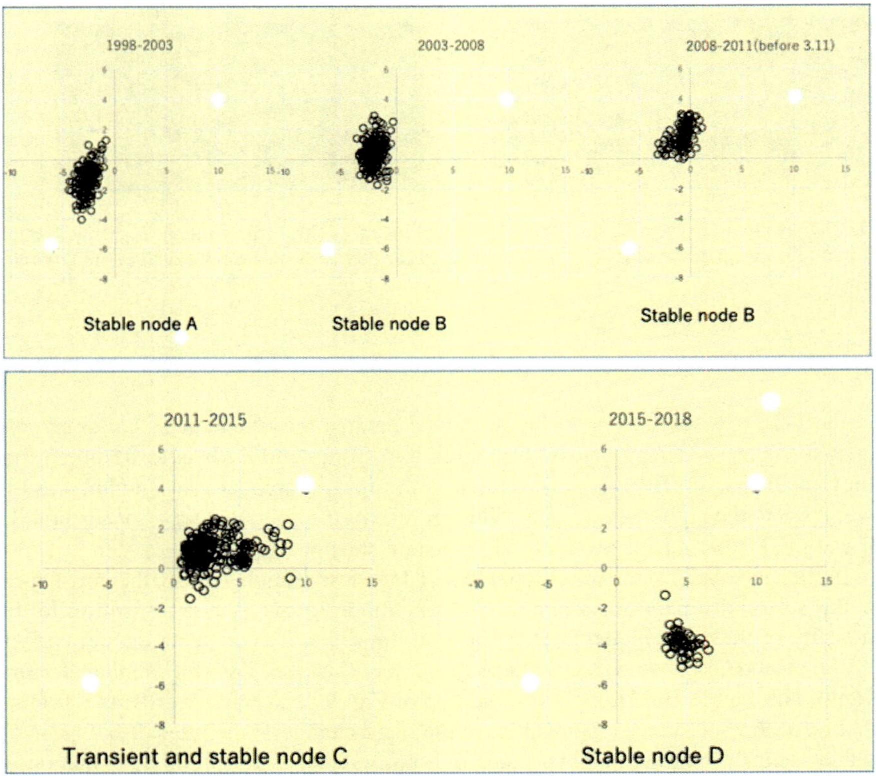

Fig. 9.8 Successive change of the stable nodes in the diagram of the correlated seismicity rates z_1 and z_2 before 2011 (top) and after 2011 (bottom) in the Japanese islands region

9.3 Partial b-Value Change and Its Annual Variation

The partial b-value changes of b123 and b234 are discussed in the previous chapters, and it is already shown that those in the mantle is likely to those in the crust in terms of frequency and the phases. Thus, it may conclude that the reason of the cyclic variations of them is the periodic activity of the slow slip events along on the plate boundary. Schwartz and Rokosky (2006) summarized the periodicity of the SSEs in the global plate boundary zones and obtained the result that the SSEs occur in the time scales of subannual to several year intermittency in the depth around 10–50 km of the subduction plate boundary (Ito et al. 2007). In additions, the SSEs of the Cascadia plate boundary occur in the intervals of 1.5 years, and they commonly associated with non-volcanic tremors at the intervals (Kao et al. 2006). Thus, it is sure that the SSEs activate the seismicity around the plate boundary both in the crust and the plate interior (Nakajima and Uchida 2018), and it may be responsible for the increase of elastic strain magnitude or the pore water content surrounding the SSE

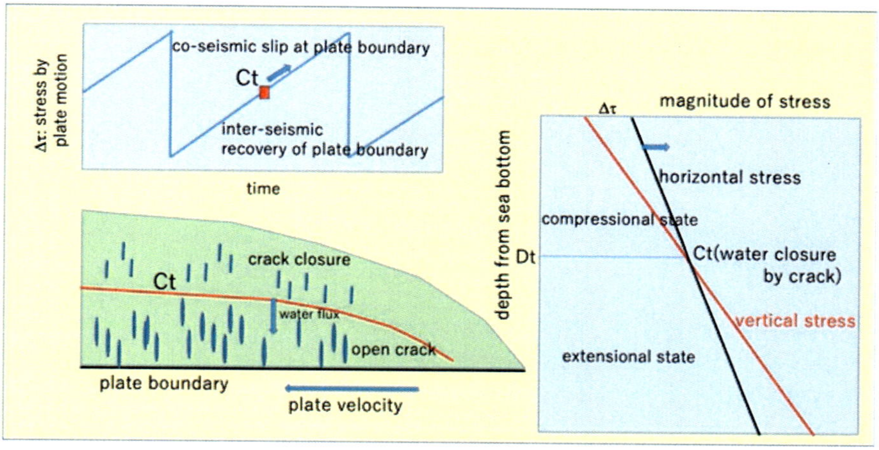

Fig. 9.9 Illustrations of the closure and opening of the small open cracks in the hanging wall crust over the subducting slab. Bottom figure shows the boundary (Ct; red line) of the closure and opening of type I cracks by the change of stress configuration with decoupling/coupling of slab and above crust as indicated in right figure. Top figure shows the differential stress changes periodically with increasing depth of Ct

plane. Considering the dehydration reaction of the plate boundary rocks subducted into the high-pressure and high-temperature conditions, that is metamorphism of the plate boundary rocks, the pore water content should increase in the rocks of the surrounding regions of the plate boundary together with the increase of strain magnitude producing the pores filled with water as illustrated in Fig. 9.9.

Partial b-value change is probably derived from the percolation of the shear cracks in the rocks system. If there are abundant shear cracks in the rock mass and the size of their shear plane varies l to m, corresponding to the earthquake magnitude, the population number of the shear cracks of magnitude l to be $N(l)$ is related to $N(l + 1)$ of shear cracks corresponding to the magnitude $l + 1$ with the partial b-value as follows:

$$\begin{aligned} -b &= \log(1 + N(l+1)) - \log(1 + N(l)) \\ &\sim \log N(l+1) - \log N(l) \end{aligned} \quad (9.12)$$

On the other hand, the order parameter in the percolation dynamics is defined by the following

$$s = N(l+1)/(N(l) + N(l+1)) \quad (9.13)$$

$$\sim N(l+1)/N(l) \quad (9.14)$$

where s satisfies $0 < s < 1$.

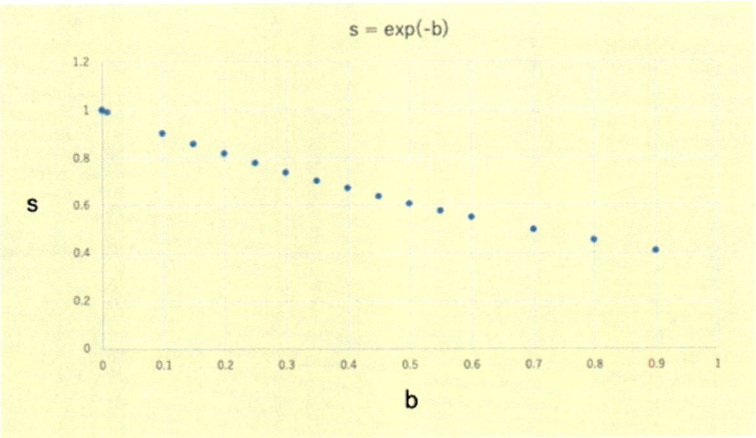

Fig. 9.10 Simplified relation of percolation parameter s and partial b-value (see text)

Combining two equations, the partial b-value holds

$$s = \exp(-b) \tag{9.15}$$

Therefore, thinking that when the order parameter of the percolation of the large size shear crack exceeds the critical value, the percolation transition yielding the total slippage should take place, and the partial b-value is a potential critical indicator for the total rupture of the rock mass.

In Fig. 9.10, it seems that the partial b-value changes gradually from A to B clusters, and at the same time, the variation of b-value increases toward the high b-value, and then, it jumps to the C cluster, suggesting that the order parameter of percolation of large shear cracks may appear increasing (1998 to 2010) and then changes to high value to giant earthquake generation (2011). After 2011/3/11, the order parameter decreases gradually to the past level as shown in Fig. 9.9.

The short-term periodic variation of the correlated partial b-value of the crust and mantle is also resulted from the cyclic change of the percolation order parameter of different sized shear cracks generation. It seems that the cyclic change of percolation order parameter is responsible for the repeated coagulation and breakout of the small shear cracks to change the partial b-value. It may associate with increase in the volume ratio of pore fluid along the shear crack interstice, and induce the low frequency earthquakes such as non-volcanic tremors. Such repeated nature of the small shear crack population related with pore fluid migration and concentration should be controlled by the annual or near annual slow slip motion along on the shallow plate boundary that yields large elastic and plastic strains together with micro-pore generation and fluid migration from dehydration along the plate boundary. Recent study by Nakajima and Uchida (2018) clarified that the seismic activity related to the dehydration of the subducted plate results the fluid upward migration within the wedge mantle,

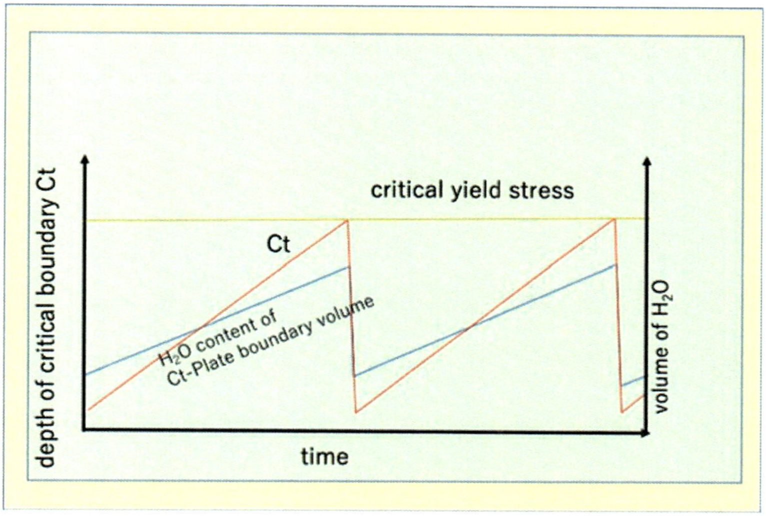

Fig. 9.11 Schematic diagram of the change in depth of Ct (red) and H_2O content (blue) in wedge crust with time after decoupling of slab and crust at the plate boundary by large slip

and then, it induces the seismic activity of the wedge mantle. It follows that the low-frequency earthquake clusters in the subduction plate becomes probably a reason of the seismic clusters of the mantle and the crust sandwiching the subducting oceanic plate as shown in Fig. 9.11.

The time scales of repeating plate boundary SSE vary in the range from the subannual to several annual periods, and thus, the correlated partial b-value change studied here has various repeating periodicity mentioned above as illustrated in Fig. 9.11. Such cyclic near annual variations of the crust and mantle in the correlated partial b-value change in the Japanese islands region is reasonable to be understood by plastic and elastic strains derived from such repeating SSE associated with diffusive and permeable flow of aqueous fluid in pores. Therefore, the coherency of phase between the mantle and the crust is possibly reasonable because of coeval fluid flux and strains between them due to the cyclic SSE along on the plate boundary.

References

Ito Y, Obara K, Shiomi K, Sekine S, Hirose H (2007) Slow earthquakes coincident with episodic tremors and slow slip events. Science 315:503. https://doi.org/10.1126/scienc.1134454

Kao H, Shan S, Dragert H, Rogers G, Cassidy JF, Wang K, James T, Ramachandran K (2006) Spatial-temporal patterns of seismic tremors in northern Cascadia. J Geophys Res 111:B03309. https://doi.org/10.1029/2005JB003727

Kuramoto Y (2003) Chemical Oscillations, Waves, and Turbulence, Dover, 340 pp

Nakajima J, Uchida N (2018) Repeated drainage from megathrusts during episodic slow slip. Nat Geosci 11:351–356. https://doi.org/10.1038/s41561-018-0090-z

Schwartz S, Rokosky JM (2006) Slow slip events and seismic tremor at circum-Pacific subduction zones, 2007. Rev Geophys 45, RG3004. https://doi.org/10.1029/2006rg000208

Chapter 10
Prediction Modeling of Global and Regional Seismicity Rates

Abstract The correlated seismicity is basically the newly defined time series data transformed uniquely by the rotation basis orthogonal matrix with assumption of sparsity and Gaussian-type noises. The time series of the correlated seismicity is formulated by the state-space modeling method to their future behavior, that is the prediction of the seismicity. Furthermore, the local seismicity rates both in the global and regional cases may be estimated with probability by means of inversion method from the correlated seismicity rates.

Keywords Ten-step-ahead prediction of correlated seismicity · Inversion of seismicity using the main sequence of correlated and raw seismicity rates · System model of subduction zone for periodic change in correlated seismicity

10.1 State-Space Modeling of the Global and Japanese Seismicity Dynamics

According to the interpretations and discussions concerning with the global and regional correlated seismicity rates features mentioned previously, it is ready to discuss the quantitative modeling of the global and Japanese islands region correlated seismicity rate dynamics to predict the near future of the seismicity. There should be two ways for quantitative modeling of global and regional correlated seismicity: One is the recurrent neural network modeling with deep learning, and the other is the state-space modeling (SSM) with Kalman filtering and smoothing of time series by Kitagawa (2010).

As described in the earlier chapter (Fig. 4.2), the state-space modeling of the time series of Helske (2018) can be formulated by the follows;

Electronic supplementary material The online version of this chapter (https://doi.org/10.1007/978-981-15-5109-3_10) contains supplementary material, which is available to authorized users.

M. Toriumi, *Global Seismicity Dynamics and Data-Driven Science*, Advances in Geological Science,
https://doi.org/10.1007/978-981-15-5109-3_10

$$y_t = F_t x_t + \varepsilon_t \quad \text{(observation equation)}$$
$$x_{t+1} = H_t x_t + R_t \sigma_t \quad \text{(state equation)} \tag{10.1}$$

where y and x vectors are observable and hidden state variables, respectively. F, H, and R are the system matrix dependent on the model, and the ε and σ are the Gaussian noises of the observation with $N(0, \varepsilon^2)$ and state with $N(0, \sigma^2)$ processes, respectively. In the state-equation model for random walk process with drift, the formulation holds

$$u_{t+1} = u_t + n_t + \sigma_t^2$$
$$y_t = N(u_t, \sigma^2) \quad \text{(Gaussian random walk)}, \tag{10.2}$$

in which u_t and n_t are random position and drift at a time t, $N(u_t, \sigma^2)$ is the Gaussian distribution function with mean of u_t and variance of σ^2. The σ_t is the Gaussian noise of the state equation.

Then, system matrixes are follows;

$$F = (1, 0), \quad H = \begin{pmatrix} 1 & 1 \\ 0 & 1 \end{pmatrix}, \quad u_t = (u_t, n_t)^T, \quad R_t = (1, 0)^T,$$
$$n_t = n \quad \text{(constant)}, \tag{10.3}$$

in which suffix, T means the transposed matrix (vector).

The correlated seismicity rates z_1 are approximated to be a noisy time series with some jumps due to large earthquakes, and those of z_2 and z_3 are cyclic time series with Gaussian noises, and in addition, the z_4 is possibly approximated to be random noise of Gaussian type. Therefore, it is reasonable that the z_1, z_2, and z_3 are to be processed independently by the state-space modeling mentioned above. After filtering and smoothing the raw data of z_1 to z_3 by SSM method, the model system composed of z_1 to z_3 is able to be obtained as the minimal dynamics model of the correlated seismicity rates in the global and regional system.

The secular variation of z_2 and z_3 shows cyclic change, and thus, the observation equation may have the following structure to be decomposed.

$$y_t = u_t + c_t + w_t \tag{10.4}$$

in which c_t is the periodic components of the z_2 and z_3, and u_t is their secular trend.

Here, the simple state-space modeling with Gaussian process will be applied for the analyses of z_1 to z_3 with random z_4 noise. The procedure of application program of state-space modeling is followed by KFSM packages in R on RStudio according to the guidance of Helske (2018).

Using the time series of z_2 to z_3 of the Japanese islands region and global plate boundary zones, the observation and state variances of Gaussian noise are given as 9.5 and 4.3, respectively. The initial conditions are applied as

$$y_1 = a_1 = (0, 0), \quad Cv_1 = \begin{pmatrix} 0 & 0 \\ 0 & 0 \end{pmatrix}, \quad Cv_{inf,1} = \begin{pmatrix} 1 & 0 \\ 0 & 1 \end{pmatrix}, \tag{10.5}$$

here, Cv is the covariance matrix of the u and c of above equation.

Then, the step-by-step smoothing and filtering curves of the correlated seismicity rates z_1 to z_3 by KFSM are estimated to be shown as Fig. 10.1 for the Japanese islands region. The smoothing curves of z_1 to z_3 by Kalman filtering are very likely to those of migration averages with 5–10 steps which means 50–100 days, suggesting that the variances of 5–10 steps in time series do not show large fluctuation. Step-by-step SSM smoothing and predictions by the above method also have been carried out for the time series of global and Northern California correlated seismicity rates z_1 to z_3 as shown in Figs. 10.2 and 10.3, respectively. The predictions of the sets of z_1, z_2, and z_3 for these cases can be estimated within the 80% reliability range for the data until ten-step forward as shown in Figs. 10.4, 10.5, and 10.6, respectively. The predictions are comparably suitable for estimation of the near-future correlated seismicity rates.

Using the one-step-ahead prediction algorithm of the KFAS modeling, the stochastic predictions until 100-step forward of the z_1, z_2, and z_3 are possibly plotted in the diagrams of z_1–z_2 and z_2–z_3 (Figs. 10.7, 10.8, 10.9, and 10.10). As shown in these figures of the correlated seismicity rate diagrams, the predicted z_1, z_2, and z_3 of the Japanese islands region (Fig. 10.7) are plotted in the cluster D belonging the recent seismic activity from 2016. On the other hand, the correlated seismicity diagram in Fig. 10.8 shows the plots of predicted z_1, z_2, and z_3 in the case of the global subduction zone within the cluster of B rather than C, and it seems that the seismic activity shows a comparably stable state (Fig. 10.8). In the case of the global ridge zones (Fig. 10.9), the predicted z_1 and z_2 are plotted in the relative high domain,

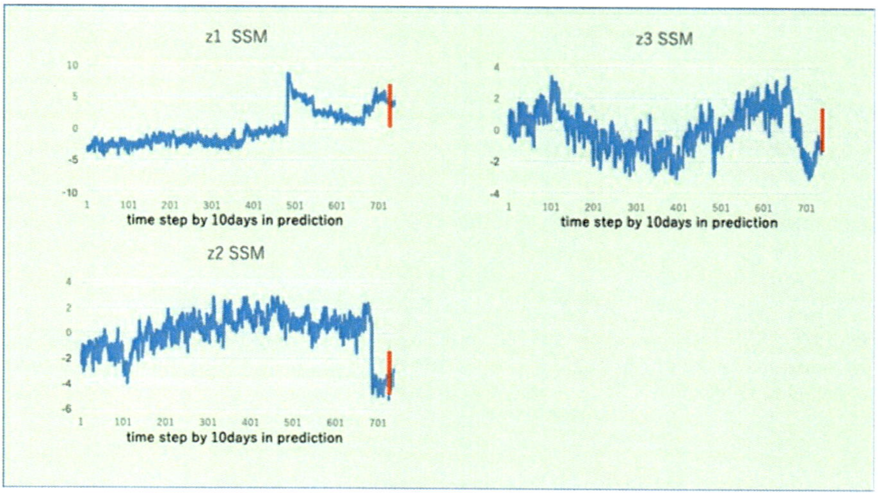

Fig. 10.1 SSM modeling of the correlated seismicity rates z_1 to z_3 by filtering and smoothing in the Japanese islands region. Red bar represents 80% reliability range of ten-step-ahead prediction

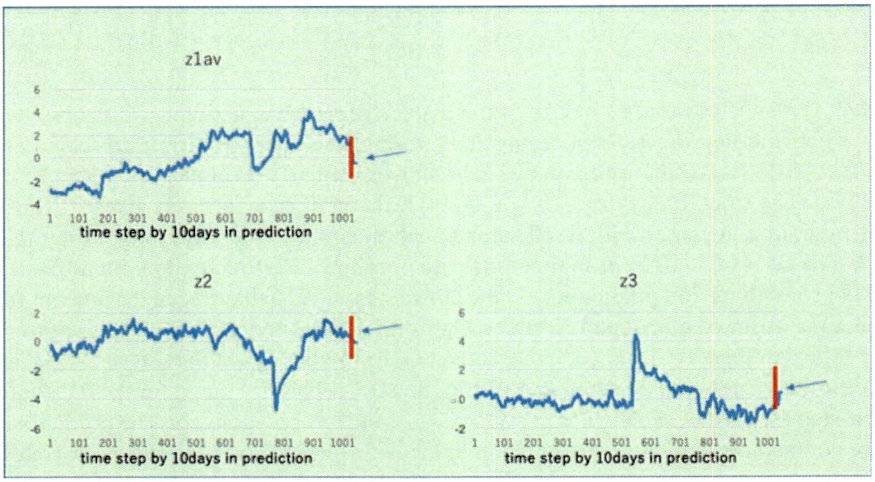

Fig. 10.2 SSM modeling of the average correlated seismicity rates z_1 to z_3 by filtering and smoothing in the global subduction zone. Red bar represents 80% reliability range of ten-step-ahead prediction

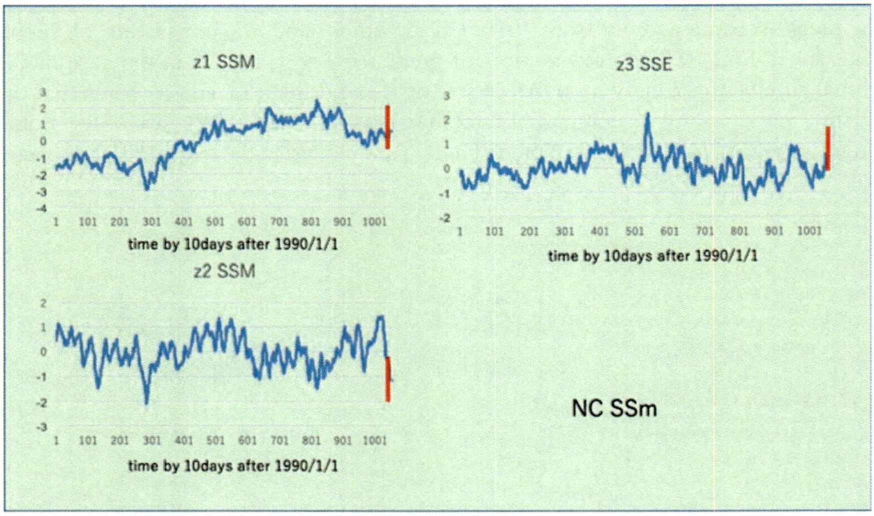

Fig. 10.3 SSM modeling of the average correlated seismicity rates z_1 to z_3 by filtering and smoothing in the Northern California region. Red bar represents 80% reliability range of ten-step-ahead prediction

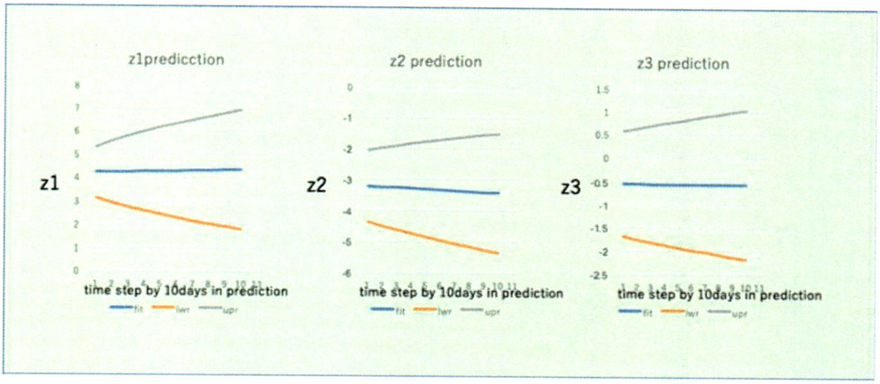

Fig. 10.4 Prediction to 100 days ahead of the correlated seismicity rates z_1 to z_3 of Japanese islands region by SSM modeling, with the range of 80% reliability

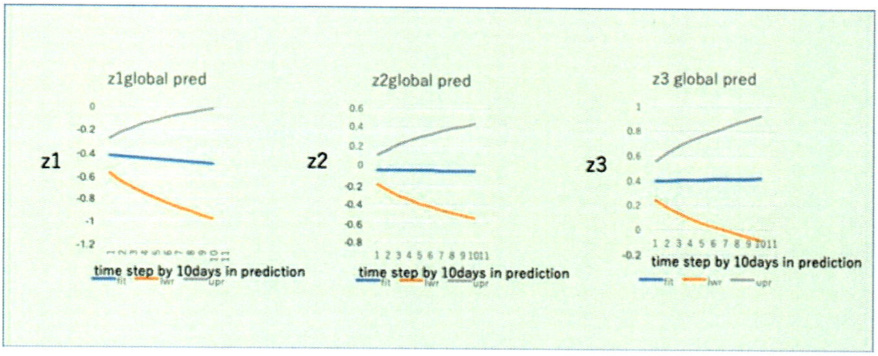

Fig. 10.5 Prediction to 100 days ahead of the correlated seismicity rates z_1 to z_3 of the global subduction zone by SSM modeling, with the range of 80% reliability

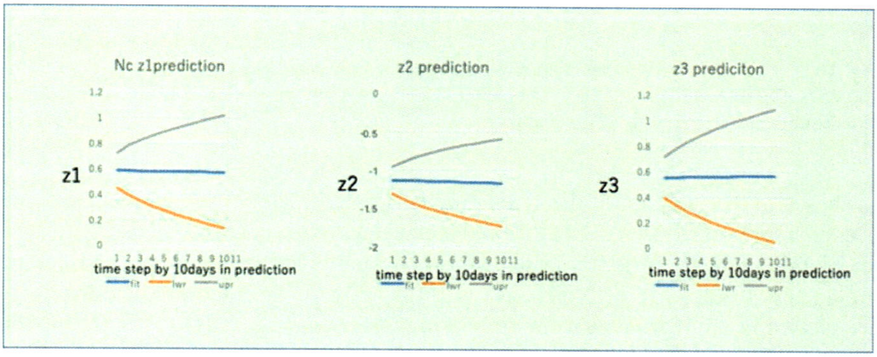

Fig. 10.6 Prediction to 100 days ahead of the correlated seismicity rates z_1 to z_3 of the Northern California region by SSM modeling, with the range of 80% reliability

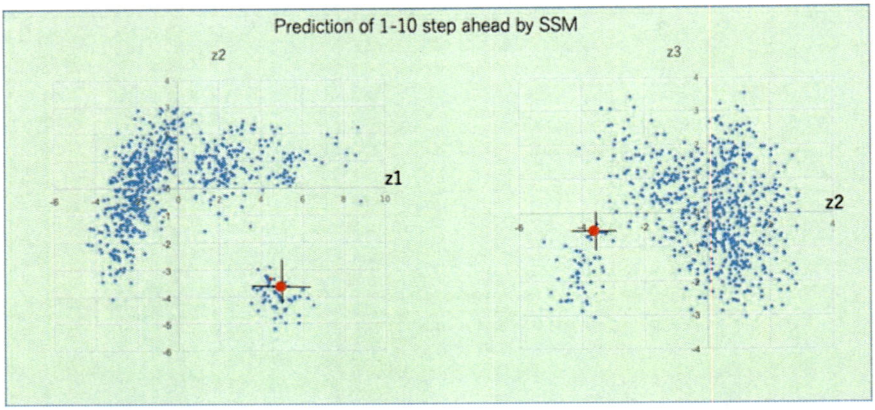

Fig. 10.7 Prediction (red circles with cross bars) of the correlated seismicity rates z_1 to z_3 to the 100 days ahead by SSM modeling in the diagrams of z_1–z_2 and z_2–z_3 with 80% reliability in the Japanese islands region with 80% reliability

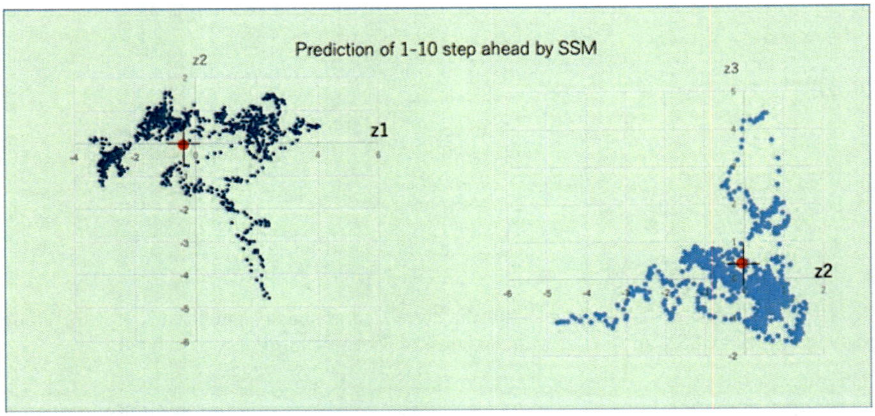

Fig. 10.8 Prediction (red circles with cross bars) of the correlated seismicity rates z_1 to z_3 to the 100 days ahead by SSM modeling in the diagrams of z_1–z_2 and z_2–z_3 with 80% reliability in the global subduction zone with 80% reliability

as shown in the diagram of the z_1 relation between the global subduction zone and oceanic ridge zone (Fig. 10.10), the predicted values of the correlated seismicity rates are plotted in the B cluster likely to the case of the global subduction zone discussed previously. In the case of Northern California (Fig. 10.11), the predicted z_1 to z_3 are plotted in the regime near the right wing of the seismic activity, suggesting the similarity of the recent activity from 2014 discussed in the precious chapter.

By the way, the local seismicity rates matrix X can be inferred by means of inversion from Z using W as follows;

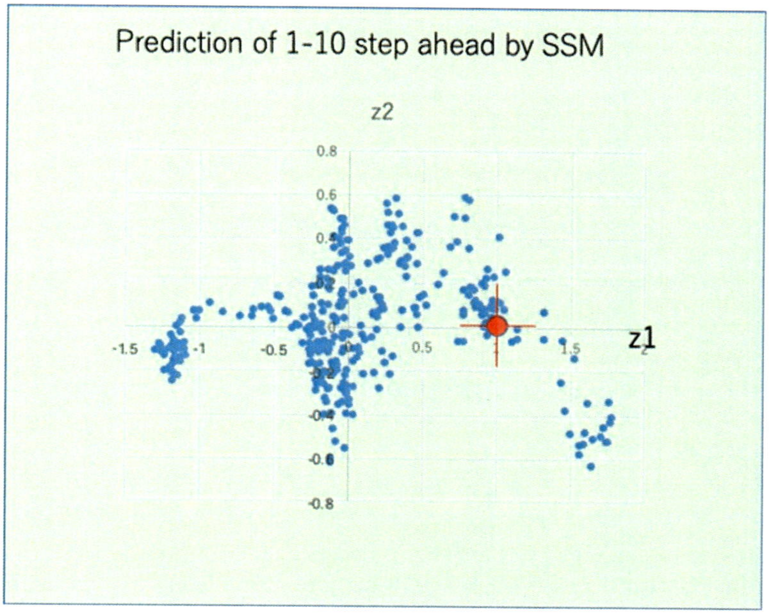

Fig. 10.9 Prediction (red circles with cross bars) of the correlated seismicity rates z_1 to z_3 to the 100 days ahead by SSM modeling in the diagrams of z_1–z_2 and z_2–z_3 with 80% reliability in the global oceanic ridge zone with 80% reliability

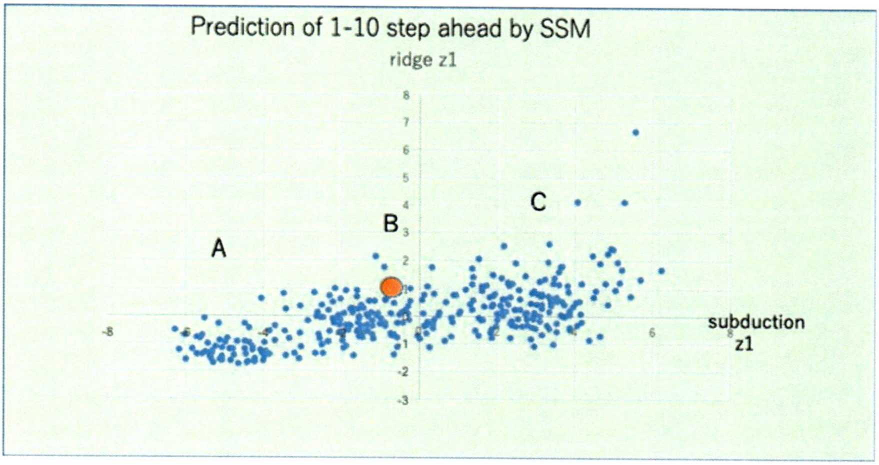

Fig. 10.10 Prediction diagram (red circles) of the correlated seismicity rates z_1 between the global subduction and oceanic ridge zone to the 100 days ahead by SSM modeling

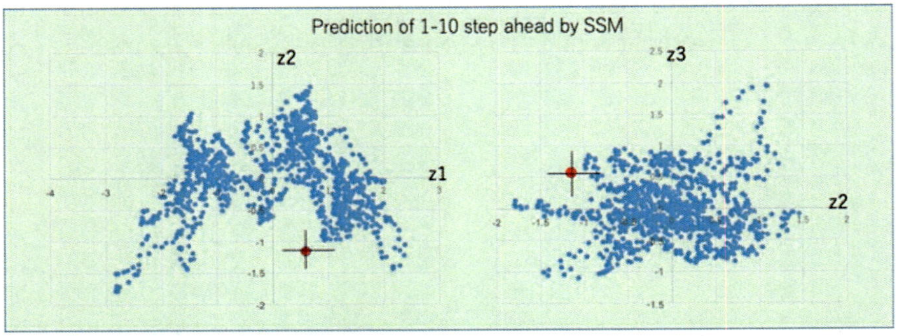

Fig. 10.11 Prediction (red circles with cross-bars) of the correlated seismicity rates z_1 to z_3 to the 100 days ahead by SSM modeling in the diagrams of z_1–z_2 and z_2–z_3 in the Northern California region with 80% reliability

$$Z = WX + \text{Noise} \tag{10.6}$$

However, the correlated seismicity z_k is degenerated into three dimensions or less and W matrix also shrink its rank = 3. Therefore, it needs that $x_j(t)$ is inferred by means of stochastic method using the singular value decomposition of W to $\mathrm{VSU^T}$, where V is the orthogonal matrix in z space and U is that in x space, and S is diagonal matrix.

Then, above equation becomes,

$$\begin{aligned} Z &= WX + N \\ &= \mathrm{VSU}^T X + N \end{aligned} \tag{10.7}$$

And

$$\begin{aligned} X &= W^T \left(W W^T\right)^{-1} (Z - N). \\ &= K(Z - N). \end{aligned} \tag{10.8}$$

In these matrix equations, X is $1 \times p$ matrix, and Z is $1 \times q$ matrix with $p > q$, and then W is $q \times p$ matrix. Then, $W^T W$ is $q \times q$ regular matrix. K is so-called pseudoinverse matrix.

In the previous studies, the W matrix is inferred by means of sparse modeling of PCA (sparsepca in R), and thus, the pseudoinverse matrix is possibly calculated as follows;

$$K = q \times p \text{ matrix} \tag{10.9}$$

In which z_1 to z_3 are obtained already as the time series by means of the state-space modeling method (SSM). Therefore, it concludes that $x(i)$ can be obtained as the time series which is possibly extended into the future cases; that is the prediction of the local seismicity rates both in the global and regional cases.

It is important that all local seismicity rates cannot be inferred from K matrix with time series of z_i because of the dimension reduction of global and regional seismicity dynamics systems; in other words, the seismicity dynamics may be described in terms of stochastic dynamical system with phase reduction equations by the sparse modeling.

This can be interpreted by thermodynamics of the continuum system such as elastic mass: In the solid elastic mass, the macroscopic parameters can be defined as temperature, pressure, elastic moduli, density, and vacancy density, permeability, susceptibility, and resistivity. However, all individual atomic displacements, atomic oscillation, lattice vibration, and other local parameters are needed for clarifying the detailed dynamics of the mass. The former is the macroscopic parameters in the nearly infinite number of elements, but in the system of huge number of elements, if the stochastic mechanical probability density function (distribution function) is possibly calculated, the macroscopic parameters can be estimated at the given boundary conditions. In this book, the z_i are considered to correspond to those of the huge number of seismic data in the global and regional solid earth.

10.2 Inversion of the Global Seismicity Rates from Correlated Seismicity

In the case of the global seismic dynamics involving subduction boundary and spreading boundary zones, the major components of every x_i are represented by the nodes of several active seismicity networks as shown in Fig. 8.11. The common locality cells are NJ, SM, TK, SS, and CH, and other cells are subsidiary. Therefore, it seems that the predicted seismicity rates of each major cells are possibly calculated by the linear combination of z_j ($j = 1, \ldots, 4$) as following;

$$x_i(t) = \Sigma g_{ij}\, z_j \quad \text{for } i = 1, 2, \ldots, 5 \text{ and } j = 1, 2, \ldots, 4. \tag{10.10}$$

where z_j involves Gaussian noises, and summation is made by j. These equations have four independent equations and others are dependent.

The relations above indicate that the locality cells of NJ, SM, TK, CH, and SS are presumably the key locality of the global seismicity dynamic system, and others may be subsidiary dependent on the activity of these key localities. The transition discussed previously from the Indian rim to Pacific rim are probably controlled by the coupling mode of their locality cells seismicity. As shown in the figure shown previously, it seems that the transition between them appears gradually in the z_1 and z_2 diagram, but their intensities of correlated seismicity becomes weak and comparable to the intensity of the Gaussian noise. It suggests that the transition of above coupling mode is likely to that of the Gaussian network dynamics.

Considering the continuous change of z_1 to z_3, the inversion from the correlated seismicity change to the active seismicity of the locality cells can be easily carried out by simple arithmetic calculations by independent predicted time series of z_1 to z_3 by means of state-space modeling method likely to the case of regional dynamic case of Japanese islands. The procedures are following:

$$\begin{aligned} X &= W^t \left(W W^t\right)^{-1} (Z - N) \\ &= K(Z - N). \end{aligned} \tag{10.11}$$

where Z is 3×1 vector and X is 4×1 vector, and W is 3×4 matrix. Thus, the K matrix is the pseudoinverse matrix of 4×3 likely to those of the Japanese and global seismicity dynamics.

For the sake of the prediction of future seismicity of the locality cells, the correlated seismicity rates z_i should be extrapolated to the near future by means of the state-space modeling. As discussed in the previous sections, the state-space modeling (SSM) of the z_i time series can be performed using the Kalman filtering and smoothing with one-step ahead prediction in KFAS packages in R (Figs. 10.7, 10.8, 10.9, and 10.10). Applying the Gaussian distribution of observation and state equations, the prediction until the next ten steps (100 days future) will be estimated for z_1 to z_3 plotted in the figure mentioned above, suggesting that the seismic activities of these regions seem to keep in high levels in the correlated seismicity rates.

10.3 Data-Driven Science and Machine Learning of Global Seismicity

In the preceding chapters, the data-driven sciences of huge amounts of seismicity rate data in the global and regional cases have been discussed using important databases of USGS, ERI, UC Berkeley, NIED, IRIS, and FDSN. These databases are now becoming large in storing and accumulating the seismic data involving waveform data and source parameters data. Furthermore, in order to predict more precisely the giant earthquake in the near future, the ocean-bottom seismic network systems and on-land dense seismic network systems are now going on foundation in the vicinity of the dangerous seismic regions such as Nankai Trough, Japan Trench, Cascadia, New Zealand, California, and other regions. In the Nankai Trough region, the DONET system has founded along on the fore-arc basin along the Nankai Trough for monitoring the seismic activity and hydro-pressure variation in order to obtain the early stage signals of plate boundary earthquake. However, it is needed to investigate the predictable model for the seismicity in the plate boundary regions that from the huge amounts of seismic data accumulating every day feature patterns of many seismicity rates time series in the global and regional scales should be manifested. In order to obtain the characteristic features of the time series of the seismicity in various localities, the big data sciences that is so-called data-driven sciences relevant to the global and regional seismicity should be tried to apply for solid earth mechanics.

Data-driven sciences in solid mechanics and solid earth mechanics are now developing science to understand the stochastic behavior of the fracture spikes (earthquakes) in rocks and rock masses. There are two schemes of data-driven sciences: One is data assimilation and parameter fitting with physical model-supervised data science, and the other is unsupervised data sciences extracting the macroscopic stochastic parameters that are the feature patterns in the unknown natural system. The former involves Bayesian inversion for parameter fitting of known model, and the latter does sparse modeling using stochastic SVD and LASSO constraints. The third way for data sciences is the neural network method for discrimination and prediction problems with machine learning and state-space modeling method for prediction problem of general multivariate time series. In this book, the author already introduced the unsupervised data sciences and state-space modeling (SSM) of the global seismicity dynamics, but he also considers that the recurrent-type neural network modeling (RNN; Bishop 2006) may be useful to investigate the predictive model of the earthquakes in the plate boundary and intraplate regions.

The recurrent neural network (RNN) modeling to investigate the general multivariate time series is based on the multiple layers of many nodes flowing in and out the data matrix, and the activation function of nodes is connected zero and unity by linear (Relu), hyperbolic tangent, and logistic types. In this case, the input data matrix is generally treated as the same rank of the output data matrix, and the evaluation function is taken as the sum of square of difference between them. However, the RNN modeling for application to the predictive seismic dynamics, it is difficult to be taken into consideration that all of the seismic time series of locality cells are used to the input data matrix and the target data matrix because of irregular noises of their individual time series. Therefore, in order to avoid this type of difficulty, it should be available that the input data matrix to the RNN model is taken as the correlated seismicity rates z_i for training and test data, and then the seismicity rates x_k of the kth locality cell may be predicted by their combinations. The transformation operator from regional correlated seismicity z_i to the individual seismicity rate x_k seems not to be available for the modeling by recurrent neural network modeling directly.

First, we will see the possible functional relationships between the transformed correlated seismicity rates and the real seismicity rates in the Japanese islands region in order to discuss the large earthquake precursor. It is because SSM modeling involves linear and nonlinear transformations of the input data by means of the hidden state evolution and apparent observables of signals in the multivariate time series. If the linear transformation between correlated seismicity rates in the global system and the seismicity rates in the locality cells exists, the simple state-space modeling is quite available for the prediction of the seismicity rates with reliability data in the cells as like as pseudoinverse matrix method discussed previously.

Figure 10.12 shows the relationships between the correlated seismicity rates z and the local seismicity rate x_k in the case of Japanese islands region for Nankai Trough cells, NE Japan cells, and Okinawa cells, suggesting that the simple sum of

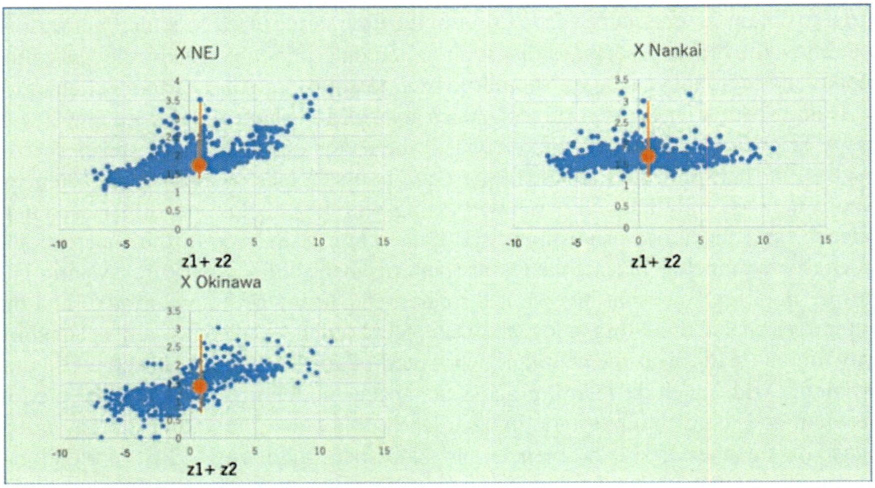

Fig. 10.12 Relation between the correlated seismicity rates $z_1 + z_2$ of the Japanese island region and the logarithmic seismicity of the locality cells of Northeast Japan (top left), Nankai (top right) and Okinawa (bottom) areas and their prediction points (red) by SSM

z_1 and z_2 is proportional to the seismicity rates of sum of x_{44} to x_{46} along the Nankai Trough. It seems that the sum of the x_{72} to x_{74} seismicity rates is considered to be also proportional to the sum of z_1 and $z_{2,}$ but there are some trends extending from the major trend showing linear relations between them.

Let us define the main sequence that is the linear relation as shown in Fig. 10.12 in the Nankai cells, NE Japan cells, and Okinawa cells against the correlated seismicity rate $z_1 + z_2$ in the region of Japanese islands. For the case of these main sequences, the simple inversion predicting the future seismicity of representative locality cells should be applicable using the prediction of the future correlated seismicity rates z_k by means of the state-space modeling method as proposed in the previous chapters instead of the DRNN modeling.

In this simple linear regression model, the error estimation is also easy to evaluate using the Gaussian noise involved in the z_k and x_i time series. The slopes of the linear main sequence defined above are 0.02 and 0.08 for the Nankai and the NE Japan region with variances of 0.4–0.5. The different relation of the signals of post- and pre-earthquake seismicity appears to be parallel to the main sequence with higher level of seismicity itself as seen in Fig. 10.12. The above relation runs along the upper limit of the main sequence with variances about 0.45, suggesting a possibility of discrimination of the post- and pre-earthquake seismicity from the main sequence of seismicity.

The main sequence of the seismicity means the background relation between the regionally correlated seismic activity and the local seismicity rates, and it should vary in the locality cells. In other words, the background relation of the Nankai Trough has smaller slope than that of the NE Japan. Therefore, it may be called as the characteristic parameter connected between the regional and the local seismic activity. Then, the characteristic parameter seems to be similar in the seismic network defined by the important eigenvector components for each correlated seismicity z_k. Remember the z_1 corresponding the seismic network mainly of Okinawa and NE Japan, the main sequence of seismicity rates of Okinawa should be likely to that of the NE Japan. Figure 10.12 indicates the relation of the sum of correlated seismicity rates, and the seismicity rates of Okinawa regions show the main sequence likely to that of the NE Japan as suggested previously.

10.4 The Main Sequence of Relations Between Global Correlated Seismicity Rates and Local Seismicity Rates: The Cases of Japanese Islands, Sumatra and Chile

In the previous section, the main sequence is possibly defined in the diagram of sum of correlated seismicity rate and the local seismicity rate, and it is approximated to the linear pseudoinverse relationship (matrix) to the real seismicity data from the correlated seismicity matrix in the case of the Japanese islands region. Next, it is important whether there is a similar main sequence in the case of the global seismicity matrix or not, because the possible pseudoinverse matrix may be a potential predictor of the future global seismic activity.

It shows that the main sequences of the global correlated seismicity rates and the local seismicity rates such as NE Japan, Sumatra, and Chile are also identical to have positive slope and linear relations. In the case of global correlated seismicity rates and the NEJ local seismicity rate, it seems that the main sequence is associated with halo extending from the main sequence and that halo is due to the large plate boundary earthquakes occurred in the NEJ as same as halo in the Japanese region case. The slope of the regression result of the main sequence is approximated to be 0.067, being smaller than that of the Japanese region case, suggesting that the local seismicity rate of the Chile as well as Sumatra is weakly sensitive to the global correlated seismicity rate change (Fig. 10.13). The main sequence defined above is seemingly applicable for the relation between the global correlated seismicity rate z_1 and the large seismic moment release rates over the magnitude 7.5 in the annual mean approximation during 1990–2018 as shown in Fig. 10.13. In this diagram, there is clearly a general main sequence and extending trends to giant plate boundary earthquakes 2004 and 2011. In this diagram, the predicted z_1 is plotted in the main sequence as shown by blue circle.

In the diagram of the main sequence of Fig. 10.14, the trend of the case for the Northern California region and also the predicted z value is shown, indicating the applicable for the general main sequence in the case of Northern California region.

Accordingly, it is reasonable that the main sequence of the correlated seismicity rates and local seismicity rates is a potential linear function corresponding to the

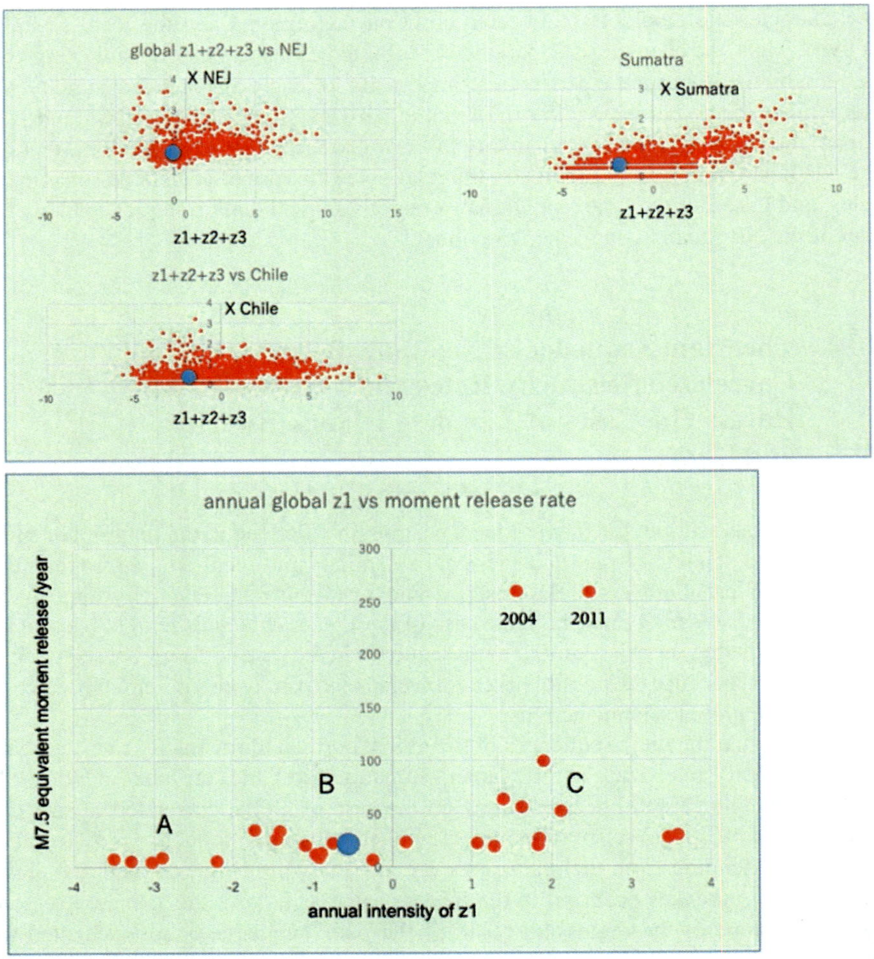

Fig. 10.13 Relation between the correlated seismicity rates $z_1 + z_2 + z_3$ of the global subduction zone and the logarithmic seismicity of the locality cells of Northeast Japan (top left), Sumatra (top right), and Chile (intermediate) areas and their prediction points by SSM. Bottom; relation between the annual intensity of the correlated seismicity rate z_1 of the global subduction zone and the annual moment release rate of M7.5 equivalent value and the prediction point (blue spot) by SSM

pseudoinverse transformation from the correlated to the local one. Then, the pseudoinverse function may be represented possibly by the SSM modeling method and by the direct linear regression method. Namely, first step is the primary data matrix transformed by logarithmic function. And then, correlated seismicity rates matrix is calculated by means of deterministic PCA or singular-valued decomposition method, or by the sparse PCA with LASSO. Successively, the smoothing and one-step prediction by Kalman filtering and prediction of future correlated seismicity matrix are

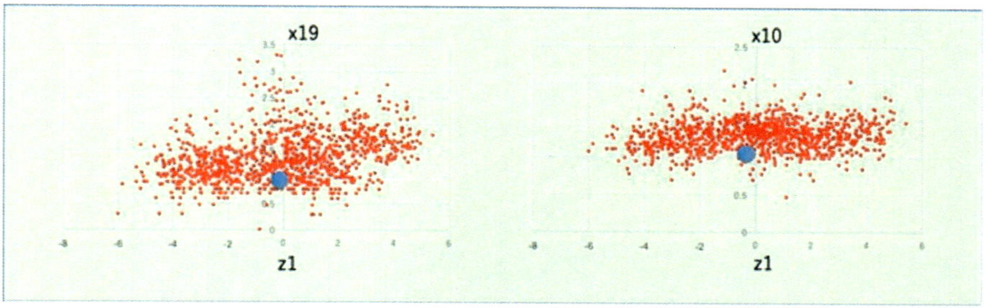

Fig. 10.14 Relation between the correlated seismicity rates z_1 of the Northern California region and the logarithmic seismicity of the locality cells of Long Valley Crater area (top left) and Geyser Geothermal Region (top right) and their prediction points (blue) to the 100 days ahead by SSM

performed by means of the state-space modeling (SSM). Predicted correlated matrix should be applied for the prediction of the real seismicity rates of the local cells by means of the main sequence obtained by the linear regression method.

10.5 Global Seismicity Dynamics and Plate Tectonics

After first of the last century, there were many giant earthquakes in the world; earthquakes over magnitude of 8.5 were counted about 16, and especially the earthquakes over M9.0 are Kamchatka 1952, Chile 1960, Alaska 1964, Sumatra 2004, and Tohoku-Oki 2011. These giant earthquakes are all plate boundary earthquake along the subduction zone. All of these earthquakes took for human societies in deep huge disaster by their giant land shake and tsunami wave. In the case of the Sumatra giant earthquake, the huge tsunami attacked the coast areas surrounding the Indian Ocean and over 300,000 peoples have died (Satake and Atwater 2007).

In Fig. 10.15, it seems to be apparent that most giant earthquakes higher than M8.5 occur along the subduction-type plate boundary and the frequency of the giant plate boundary earthquake is overwhelmingly high in the Chile–Peru and northern Pacific regions. Furthermore, it is probable that the giant earthquakes occur increasingly with the time and especially their number increases in the period of 1950–2019.

The framework of the plate tectonics (e.g., Minster and Jordan 1978) is based on the oceanic ridge as a divergence boundary of the plate motion and trench as a convergence boundary and transform fault as a shear displacement boundary. Therefore, the trench as a subduction zone of the plate is the high strain zone together with mid-oceanic ridge, but the former should favor the thrust-type displacement boundary, but the latter does the normal fault type one as stated until the foundation stage of the plate tectonics. As a result, the seismic activity of the global earth is concentrated along the plate boundary zones mentioned above, and the source mechanisms of them are different from the plate boundary zones. On the other hand, the relative velocities of the plates such as Pacific plate, Eurasia plate, Indo-Australia plate, North

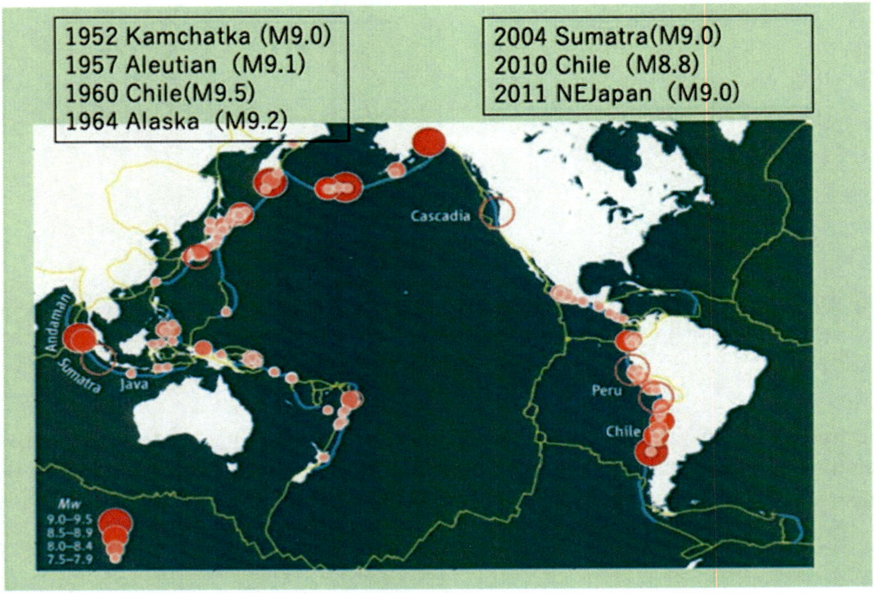

Fig. 10.15 Map showing the recent epicenters of the giant plate boundary earthquakes over magnitude 9 (personal communication by Sugioka of Kobe University)

America plate, and others are measured by means of the VLBI system (Fig. 10.16) and network GPS system: The conversion velocity along the Japan Trench reaches about 9 cm/year, and it keeps nearly constant value during 1998–2018 instead of the varying seismicity rates. Comparing the displacement variation between Hobart, Australia and Tsukuba/Kakioka, NE Japan with global z_1 time series from 2012 to 2018, there seems to be some fluctuated periods of both time series, suggesting the plate motion affecting the global correlated seismicity rate variation.

The relative displacement at the subduction plate boundary due to the plate motion is considered to be responsible partly for the boundary slip by giant earthquakes and partly for the steady-state viscous slip along the subduction boundary. Judging from the empirical relation between the slip distance and the magnitude of the earthquake proposed by Ohnaka and Matsuura (2002) and the slip distance about 70 m at the Tohoku-Oki giant earthquake (M9), average displacement rate to be about 9 cm/year which should be applied to the whole subduction zone even under the correction of rotational motion of the plate should be much larger than that of the average total displacement by large earthquakes per year. Thus, the discrepancy of them is derived from the stable viscous displacement of the slab along on the subduction boundary.

Recent network GPS studies along the subduction boundary zones clarify much of slow slip events that show the slip duration longer than several months and the slip length of about several to several ten meters. These slow slip events (SSEs) are investigated by displacement determined by GPS and gravity study but not by

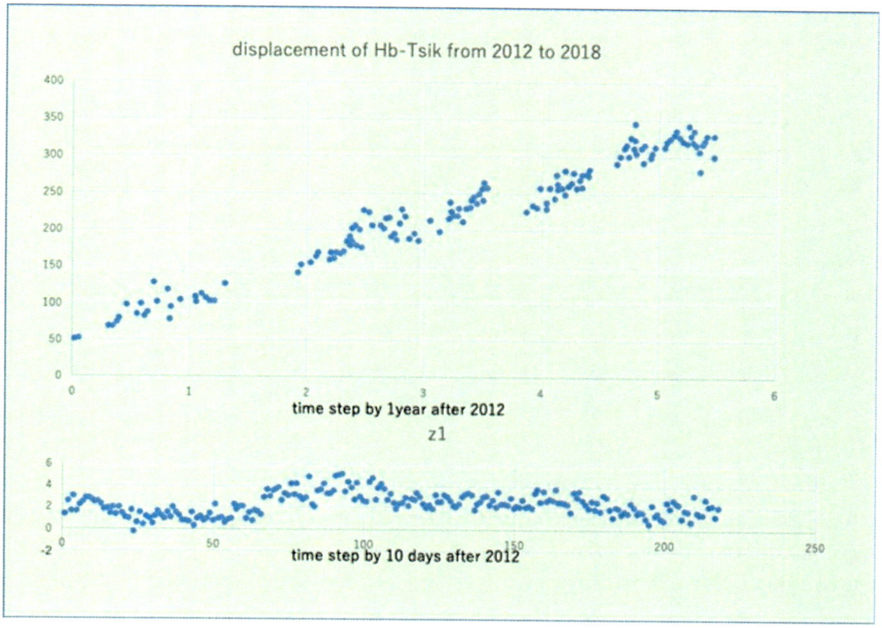

Fig. 10.16 Comparison of the correlated seismicity rate z_1 of the global subduction zone and the displacement by the VLBI method between Kakioka of Japan and Hobart of Australia

seismic wave. Furthermore, these SSEs are associated with much of micro- to small earthquakes in the surrounding volume of the crust and upper mantle. The seismic activities associated with these SSEs are characterized by the subannual to multiple annual periodic variations both in the correlated seismicity rates and in the variations of non-volcanic tremor activity as discussed in the modeling of the annual seismic variations along the subduction boundary zones of this study. Therefore, it seems to be sure that there are wide velocity variations of the stable and unstable slip between the subducting slab and overriding crust and upper mantle of the arc.

In the historical survey of the ancient large earthquakes as shown in Fig. 10.17, giant earthquakes occur frequently around the Pacific Ocean and especially their activity becomes large for the latest 120 years of 1900 AD to 2019 AD, suggesting that the plate velocity changes rapid during that period, being compared with the times from 1500 to 1900 years. In this study of the seismicity correlation method, the time scales for the investigated time series of seismicity range from 20 to 30 years, and even in this short time scales, it seems that the global and regional seismic activities are possibly assigned to be periodic in terms of the correlated seismicity rates of small to micro-earthquakes. Besides, the periodicity of the correlated seismicity rates ranges from several ten to subannual ones. It, consequently, leads that the plate velocity in the global system should be variable but not keep constant even during 20–30 years.

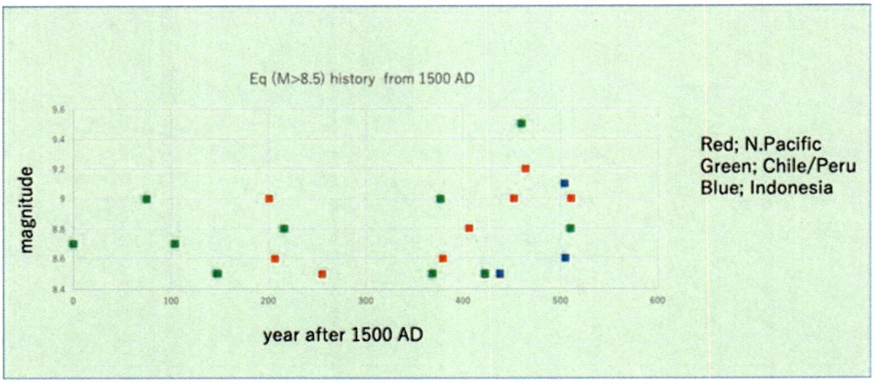

Fig. 10.17 Historical time series of the giant plate boundary earthquakes in the world from 1500 AD to recent

Considering above variation in the plate velocity in the global system, the sea level change may be also coeval with the variation of frequency in giant seismicity as mentioned above. Historical giant earthquakes have been recorded in the human community, and also the sea level change was recorded in the geological evidence such as ancient coast line and after 1990, it is possible to obtain the sea level change by satellite laser altimeter and the tidal gauge. The net global sea level change is estimated by subtraction of thermal expansion of sea water. As a result, it seems that the total amount of sea water increases continuously as a rate of 3.4 ± 0.4 mm/yr. The rate of GSL from 1990 to 2005 is rather constant, but that from 2005 to 2018 shows variable with 3–4 year periodicity. It suggests that the variation intensity of water loading and fluid pressure operated on the earth crust and mantle vary largely until 2005.

On the other hand, the historical GSL change estimated from geological evidence shows the pattern that it has small maximum at 1500 AD and becomes minimum at about 1800 AD and then it increases to 2018 AD, suggesting that the large earthquake frequency minimum at about 1800 AD looks like synchronization. This may be due to the sea water loading and fluid pressure on solid earth decrease and differential stress and effect of hydrolytic weakening become weak at the time of minimum GSL.

The global sea level rise is surely responsible for the global warming and resultant decrease of ice cap in the polar districts. Therefore, it may be provided that the global seismicity high (GSH) after 2005 is seemingly synchronized with global climate warming after nineteenth century. According to ICCP results, the global warming of the earth surface tends to continue in the near future. Thus, it may be concluded that the global seismicity high should continue extending next century. This seems to be unexpected prediction for human society and environmental crisis in the near future.

On the other hand, the global and regional correlated seismicity rates display weak but obvious periodicity around subannual to tri-annual time scales and of decadal time scales discussed in the previous chapters. Especially, their time series of z_1, z_2, and z_3 indicate the peaks at 1997, 2005, 2007, 2011, and 2014. However, the GSL change and polar ice sheet mass minimum appear at peaks of 1998, 2003, 2006, 2010, 2013, and 2016 (Chen et al. 2005; Leuliette and Willis 2011; Chanard et al. 2018). So, it is not straightforward that the short term periodicity about 3 years in the GSL and Polar Ice Mass Loss may control the correlated seismicity rate variation having several-year periodicity.

The effect of the GSL change on the solid earth is the loading and unloading stress likely to the tidal force on the earth interior. The tidal sea level change reaches about 1–2 m, and the corresponding stress change by loading and unloading becomes about 10.20 kPa which is hundred times of the GSL change mentioned above. Therefore, it may be sure that the GSL change does not control the long-term variation of the global correlated seismicity rate because of no obvious effect of tidal change of the seal level on seasonal change of correlated seismicity.

On the other hand, the GSL change is derived mainly from the melting of the polar ice sheets, and thus, the 1 cm rise of GSL corresponds to about 0.25 m thinning of total ice sheets. It may yield the unloading stress change around the high latitudinal earth interior attaining about 2.4 kPa, suggesting the ten percent of tidal loading and unloading. Therefore, the large-scale GSL and related ice sheet mass loss may not contribute mainly the periodic variation of the correlated global seismicity.

As pointed out in the previous chapters, the correlated seismicity rates of the global ridge system are strongly synchronized with those of the global subduction zones. It leads that the average strain and stress accumulation rate along the subduction zones are controlled by the spreading of the oceanic plates at the global ridges, and the stress accumulation rate should be resulted from the plate velocity at the subduction zones. Consequently, the plate convergent velocity along the subduction axes which should change according to the geometry of the rotational motion of the plate basically governs the correlated seismicity rates along the subduction zones. In order to compare the patterns between the z_1 of the global correlation seismicity and the VLBI displacement records (GSI, 2019) between Tsukuba/Kakioka in Japan and Hobart in Australia, the figures are shown from 2012 to 2018. It seems that the correlation between their patterns appears slightly in the 2013–2015, suggesting that the convergent plate velocity increases with increasing z_1 correlated seismicity even with small time lag between them.

On the other hand, there is the clear relaxation pattern of the displacement between Japan and Australia and Japan and Hawaii after 2011/3/11 Tohoku-Oki earthquake during 2011–2012 (VLBI). It is obviously due to the rapid eastward slip of NE Japan crust, being associated with viscoelastic response of the asthenosphere. Therefore, it concludes that the plate motion is largely changed by the large seismic activity variation along the plate subduction zone. Besides, it should be suggested that the change of the plate velocity is responsible for the different areas of the subduction boundary zones and also in the ridge regions. In addition, considering the viscoelastic response mentioned above, the time lag between the plate motion and the correlated seismic activity may be appeared within a few years.

As noted previously, there are several seismic active periods associated with giant plate boundary earthquakes and huge tsunami wave from 1500 to 2019 AD. The relative quiet period of large seismic activity is also found between 1750 and 1800 AD. In the sense of the long-term trend of the giant earthquake frequency, the period after 1900 AD is considered to be very active in large seismicity. Consequently, it should be probable that the plate motion in the global earth system was very active in the periods of 1500–1600 AD and of 1900–2019 AD.

By the way, the important problem exists in the relationship between the correlated seismicity rates and the occurrence of the giant earthquakes. Already, it has been discussed that the correlated seismicity rates are separated from the giant earthquake and related post-seismicity, because of the general trends connecting the clusters and different trend coming from general trend in the correlated seismicity rates diagram of z_1 versus z_2 in the global and regional systems. As shown in Fig. 9.2, the general trend is composed of the three major clusters A, B, and C, but the extension arms that display the fore- and post-seismicity of the giant earthquakes of Sumatra and Tohoku-Oki ones are extended from the cluster B in the z_1–z_2 and z_2–z_3 diagrams of the global system.

Similar relations between the main trend and the bifurcated arms are possibly found in the regional case of Japanese islands region. From Fig. 9.5, it seems that the main trends from A to C and D, but the bifurcated trend from B extending the large z_1 with constant z_2 are seen, suggesting the abnormal moment leading the giant earthquake is to be in the state in the B cluster of z_1 and z_2 diagram. In order to investigate the mechanical state of the B cluster, the fluctuation patterns of the main clusters A to D by means of comparison of annual variation diagram of z_1 and z_2 with increasing time as shown in Fig. 9.8. The annual variation of z_1 and z_2 is obtained from the subtraction of long-term secular trends of z_1 and z_2 from raw their trends. As mentioned in the previous chapter, the there are two orthogonal variations along on the z_1 and z_2 axes and the variation of z_1 rapidly becomes wide in the C cluster just before 2011 Tohoku-Oki giant earthquake with keeping the variation width of z_2. Furthermore, the variation of the z_1 of short-term fluctuation becomes very large just after 2011 in the post-seismic stage. Therefore, the variance change in the C cluster appears as the potential signal for the giant earthquake.

On the other hand, the z_2 variation of the D cluster in the period from 2016 to 2018 looks slightly different compared with the variation of the B cluster because of very large z_2 variation of the annual fluctuation. It seems that the trend of the near future change displays surely toward C from D cluster, suggesting the stable seismicity with high-level activity in Fig. 10.7.

By the way, the similar discussion may be applied to the case of the global correlated seismicity dynamics in Fig. 10.8. It is probable that the fluctuation relation between z_1 and z_2 short-term time series in the period from 1990 to 2008 appears noise like to Gaussian pattern but that in the period from 2008 to 2018 does slightly correlated feature between z_1 and z_2 short-term variations. As shown in Fig. 10.8, on the other hand, the trends of the z_1 and z_2 in the time of giant earthquakes 2004 and 2011 seem to be radiated from the cluster C but not A and B. Therefore, it concludes that when the correlated seismicity rates enter the cluster C the giant earthquake may be triggered in somewhere of the global subduction zone.

For application of the state-space modeling (SSM) for the future global correlated seismicity rates, the predicted z_1 and z_2 are plotted in Fig. 10.8, indicating the plots near the cluster B apart from the cluster C. It therefore, suggests that the global subduction system is now becoming stable state.

10.6 Possibility of Deep Learning Recurrent Neural Network for Prediction of the Seismic Activity

As discussed in the previous sections, the relations of the correlated seismicity rates and the actual seismicity rates of the local cells are possibly classified into the main sequence and the abnormal trends: The main sequence is expressed as the linear combination of the several terms of the correlated seismicity rates, and the abnormal trends are the different sequences from the main sequence. For simplified discussion, it should be considered that the prediction of the seismic activity coming large earthquake may be applied by the linear combination of the correlated seismicity rates time series approximated by the state-space modeling.

Here, the global and regional correlated seismicity rates are the macroscopic physical parameters in the complicated earth mechanics system presented previously, and the actual seismicity rates of the local cells of the subduction zone, oceanic ridge zone, and transform fault zone represent the local mechanics parameters surely depending upon above macroscopic parameters. Then, the data matrixes of the correlated seismicity rates $z(i, j)$ where j is the component number of the macroscopic parameters, and real seismicity rates $x(i, j)$ and the target variables are also $x(i, j)$, where j is the locality label and i is the time step are approximately represented by the linear combination of each other as discussed in the previous section. Then, the phase map shown by the diagram of z_1 versus z_2 and of z_2 versus z_3 should be represented by the continuous single curve with several stable nodes as shown in Fig. 10.8. These phase migration patterns look like the Lissajous patterns seen in the correlated displacement and stress of the circular network system of the slider blocks discussed in the previous chapter. In that slider block system, the elementary units of the blocks are connected with linear dynamics of elastic force side by side. Accordingly, it should be available that the inversion of the seismicity rates is possibly inferred by the simple combination of the correlated seismicity rates z_i in terms of the state-space modeling but not of recurrent neural network modeling with deep learning.

10.7 System Model of the Correlated Seismicity, Plate Boundary Slip, and Fluid Flux in the Subduction Zone

First let us discuss the emission of the earthquake wave from the source cracks. The problem is whether elastic wave is emitted at the time of rapid slip of the precut shear crack or at the case of slip of shear crack. In the former, it is assumed that the precut shear crack is elastically deformed under the local stress conditions. And the second model is based on the existing shear crack of unstrained state, and elastic wave emission is resulted from the unstable slippage by adding the shear stress over fracture strength. Considering that the probabilistic law of Gutenberg and Richter relation between magnitude and frequency of earthquakes in the space and time, the former model of elastic wave emission is probably favorable because of the size population of existing shear cracks of GR law.

There are abundant repeating earthquakes in the plate boundary zone as clarified by Igarashi et al. (2003). The repeating earthquakes called as repeater occur in the same position along the plate boundary and off the plate boundary. The duration of the repeating the earthquakes having same magnitude appears to be nearly dependent upon the magnitude, that is to say for constant magnitude it keeps constant periodicity. Judging from the fact that the magnitude corresponds to the shear crack size and slip length, and constant periodicity suggests the constant shear strain accumulation between the two blocks around locked shear crack during the interseismic period. It follows that the same locked shear cracks should be accumulated shear strain during the interseismic period and if effective shear stress on the present shear cracks in the rocks suddenly decreases or the effective water pressure rapidly increases and effective shear stress suddenly lowers, the shear stress derived from shear strain acting on the shear crack must overcome the fracture strength or frictional stress and then becomes unstable rapid slip to emit the elastic wave.

This model should be called as the shear crack reversal model of earthquake triggering because of the earthquake occurs at the time of reversal motion of strained shear cracks. After the giant earthquakes such as Sumatra and Tohoku-Oki earthquakes, there were the periods with 1–2 years of the many normal fault-type earthquakes in the plate boundary regions. It means that the shear strained and locked shear cracks might be reactivated in the change of the stress configuration in the plate boundary zone by means of the large-scale slip due to giant plate boundary earthquakes. Of cause the giant plate boundary earthquakes should be occurred by the shear stress accumulation followed by continuous plate subduction.

On the other hand, the migration of fluid such as free water in rocks should lower the friction of the shear crack surface by means of decrease of normal stress on that surface, and thus if the external force acting on that crack may overcome the internal stress derived from the displacement along on the shear crack, the shear crack rapidly takes more displacement and stop at the mechanical balance state. At this time, the elastic wave should be emitted from the shear crack. Actually, recently Nakajima and Uchida (2018) reported that the water migration related seismicity over the subduction boundary in the Nankai Trough region as discussed by Kodaira et al. (2004).

By the way, the stress configuration in the subduction zone should be changed by the large earthquakes and slow slip events along the subduction boundary because of change of mechanical coupling stress between the subducting slab and overriding arc crust. In Fig. 10.18, the possible stress configurations before and after the large plate boundary slip are shown: The maximum stress direction within the crust between

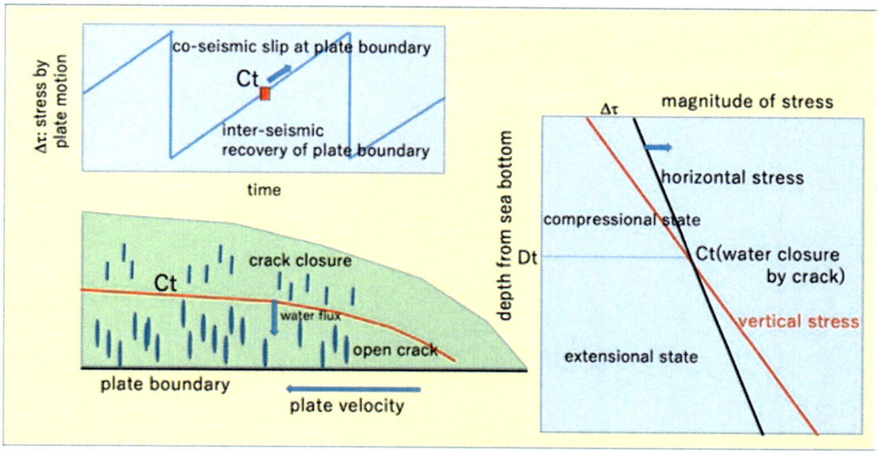

Fig. 10.18 Illustrations of the closure and opening of the small open cracks in the hanging wall crust over the subducting slab. Bottom figure shows the boundary (Ct; red line) of the closure and opening of type I cracks by the change of stress configuration with decoupling/coupling of slab and above crust as indicated in right figure. Top figure shows the differential stress changes periodically with increasing depth of Ct

the trench and volcanic arc is nearly along the plate motion in the period of strong coupling between them, but just after the slip, it should be changed to the vertical orientation by the decoupling of them. It is because the reason that the mechanical equilibrium state shows the situation that the maximum stress becomes gravitational loading stress as $\rho\ g\ h$ where h is depth and horizontal stress takes the value governed by Young's modulus (or Poisson's constant ν; Jaeger 1962) to be $\nu/(1-\nu)\sigma_1$ (Means 1976).

On the other hand, let us look at the migration of water stored in the cracks and pores in the rocks. During the co-seismic period of the large plate boundary slip, the wedge region of the crust seems to be extensional condition that yields small vertical open cracks filled with fluid and related shear cracks. After the large boundary slip, the mechanical coupling between the crust and the subducting slab becomes increasing by recovery of the slip zones and then the stress configuration of the wedge crust changes gradually to the horizontal compression state. The gradual change of this stress configuration seems to be represented by the level of vertical-to-horizontal maximum stress lowering to the plate boundary (Araki et al. 2017). It results that the fluid stored in the open cracks migrates toward the deep wedge crust over the plate boundary as shown in Fig. 10.19. As a result, fluid should accumulate in the plate boundary as well as the shear cracks of the wedge crust, reducing the effective normal stress on the shear cracks and the plate boundary. Therefore, the plate boundary slip should be reactivated by reducing the normal stress on the boundary plane due to increase of the fluid pressure.

Considering that the migration velocity of the fluid should control the time interval between the plate boundary slip, the permeability governed by the density of open

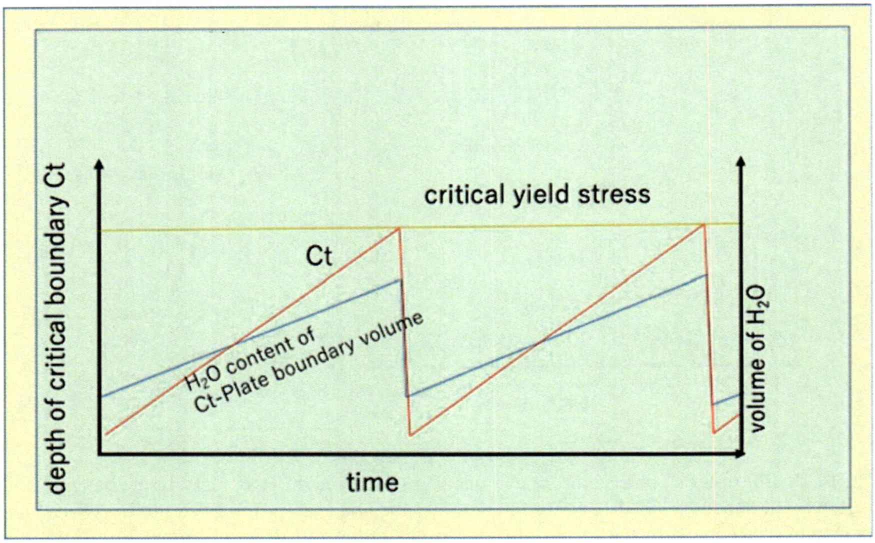

Fig. 10.19 Schematic diagram of the change in depth of Ct (red) and H_2O content (blue) in wedge crust with time after decoupling of slab and crust at the plate boundary by large slip

cracks and pores in the wedge crust is the key for the boundary slip intervals. By the way, the permeability of the rocks is governed by the porosity and the density of the cracks and the density of cracks is also dependent on the applied stress. However, the stress configuration in the wedge crust change from vertical maximum stress to the horizontal maximum stress state after major boundary slip event. Therefore, it can be imagined that the abundant open cracks filled with fluid at the time of boundary slip gradually close by the horizontal maximum stress with increasing the strength of the mechanical coupling due to recovery of the plate boundary zone. This means the lowering the effective permeability, and then, the fluid migrates into deeper zone of the wedge crust toward the plate boundary (Fig. 10.18).

If the average permeability of the rocks in the wedge crust is constant during the interseismic period, the time scale of the interseismic period may be estimated by the recovery rate of the plate boundary zone after boundary slip. Simple model is possible to consider: The migration of fluid in open cracks is controlled by the advancing rate of the closure zone of the open cracks toward the boundary which is determined by the increasing rate of the horizontal stress as shown in Fig. 10.19. Therefore, it concludes probably that the time scale of the mechanical coupling recovery of the boundary zones without slow slip motion is then approximated to be the time scale of the accumulating fluid along the boundary zone and of the inter-seismic period of large plate boundary earthquakes.

References

Araki E, Saffer DM, Kopf AJ, Wallace LM, Kimura T, Machida Y, Ide S, Davis E, IODP Ex.365 shipboard scientists (2017) Recurring and triggered slow-slip events near the trench at the Nankai Trough subduction megathrust. Science 356:1157–1160

Bishop CM (2006) Pattern recognition and machine learning. Springer, New York

Chanard K, Fleitout L, Calais E, Rebischung P, Avouac JP (2018) Toward a global horizontal and vertical elastic load deformation model derived from GRACE and GNSS station position time series. J Geophys Res Solid Earth 123(4):3225–3237

Chen JL, Wilson CR, Tapley BD, Famiglietti JS, Rodell M (2005) Seasonal global mean sea level change from satellite altimeter, GRACE, and geophysical models. J Geodesy 79(9):532. https://doi.org/10.1007/s00190-005-0005-9

Helske J (2018) KFAS: exponential family state space models in R. https://cran.r-project.org/web/packages/KFAS/vignettes/KFAS.pdf

Igarashi T, Matsuzawa T, Hasegawa A (2003) Repeating earthquakes and interplate aseismic slip in the northeastern Japan subduction zone. J Geophys Res Solid Earth 108(B5)

Jaeger JC (1962) Elasticity, fracture and flow with engineering and geological applications. Wiley, New York

Kitagawa G (2010) Introduction to time series modeling. Monographs on statistics and applied probability, vol 114. CRC Press, New York 289 pp

Kodaira S, Iidaka T, Kato A, Park J, Iwasaki T, Kaneda Y (2004) High pore fluid pressure may cause silent slip in the Nankai Trough. Science 304:1295–1299. https://doi.org/10.1126/science.1096535

Leuliette EW, Willis JK (2011) Balancing the sea level budget. Oceanography 24:122–129

Means WD (1976) Stress and strain. Springer, New York

Minster JB, Jordan TH (1978) Present-day plate motions: a summary. In: Allegre JC (ed) Source mechanism and earthquake prediction. CNRS, Paris, pp 109–124

Nakajima J, Uchida N (2018) Repeated drainage from megathrusts during episodic slow slip. Nat Geosci 11:351–356. https://doi.org/10.1038/s41561-018-0090-z

Ohnaka M, Matsu'ura M (2002) The physics of earthquake generation. Univ. Tokyo Press, Tokyo, 378 pp (in Japanese)

Satake K, Atwater BF (2007) Long-Term perspectives on giant earthquakes and tsunamis at subduction zones. Annu Rev Earth Planet Sci 35(1):349–374

Chapter 11
Problems of Prediction for Giant Plate Boundary Earthquakes

Abstract Even from the micro- to small earthquakes activity in the global earth, the prediction of the large to giant earthquakes may be very difficult because of their rare occurrence. However, the general relationships between the seismically active regions and the correlated seismicity are possibly potential indicator for the bifurcation to the large and giant earthquake trends, though they are different with every region.

Keywords Giant earthquake · Prediction by big data-driven science · Future seismic network system

It is quite important that the short- and long-term prediction of the giant boundary earthquakes along the subduction zones makes possible to realize and apply for the human system, because of many big cities of the human society in the world are enlarged along the coast area near the subduction zones. Especially, it is evident that there are many giant cities near the coast facing with the future giant tsunami disaster like 2004 Sumatra and 2011 Tohoku-Oki giant earthquakes. These giant earthquakes gave and continuously give the huge amount of damages in Japan and SW Asian countries. In order to avoid the severe damages in person and society, the predictive information on the intensity and regions of future earthquakes and related tsunami is quite needed, and it should be open in society as noted in Satake and Atwater (2007).

In this book, the author intends to try to investigate the global seismicity dynamics by means of big data-driven science and predictive work of the seismic activity in near future using the machine learning with state-space modeling and deep neural network modeling. As a result, the spatially correlated seismicity rates, these are called as the correlated seismicity, are the good indicators of the global and regional seismic activity: The periodic variations having the long-term and short-term cyclic variations are possibly identified from the random noises of Gaussian type.

Electronic supplementary material The online version of this chapter (https://doi.org/10.1007/978-981-15-5109-3_11) contains supplementary material, which is available to authorized users.

M. Toriumi, *Global Seismicity Dynamics and Data-Driven Science*, Advances in Geological Science,
https://doi.org/10.1007/978-981-15-5109-3_11

Furthermore, it is suggested that the exact inversion from the correlated seismicity rates time series to the local cell seismicity rates can be performed by the inverse matrix of orthogonal transformation basis matrix W as mentioned above with Gaussian noises term. It follows the equations of

$$x(t) = W^{-1}(z(t) + z_g(t)) \quad (11.1)$$

in which z_g is the Gaussian noise component with mean value equal to zero of the correlated seismicity rates. Therefore, the following equations hold,

$$x_k(t) = w'_{kl}z_l(t) + w'_{k2}z_2(t) + w'_{k3}z_3(t) + w'_{k4}z_4(t) + \sum \left(w'k_j z_j\right) \quad (11.2)$$

$\sum \left(w'k_j z_j\right)$ = average Gaussian noise and w' is the components of the W^{-1} matrix.

In the previous sections, the relations of the locality cell seismicity and the global and regional correlated seismicity are shown in Figs. 10.12 and 10.13, and they are approximated to the above equation with the terms of significant coefficient. The machine learning state-space modeling inversion from z and x time series is then obviously possible to take inversion because of the linear relation of them with Gaussian noise terms.

By the way, the probabilistic prediction for the large earthquakes involves several terms: The first is the date of the earthquake, the second is the locality, the third is the magnitude, and the fourth is the related tsunami (Ogata 2015). These sets of information about especially giant earthquake are needed for avoiding the severe disaster by earthquakes. In this book, the author intends to propose a dynamics and inversion of the activity of micro- to small seismicity rates in the locality cell of the regional and global system, but cannot suggest the clear method to predict the large earthquake probability itself because there is still unclear evidence on the relationship between the micro- and small seismic activity and the emission of the large earthquakes.

On the other hand, the information of prediction of large earthquakes is different from that of weather forecast common in the broadcast system: In the case of weather forecasting, even extreme weather such as heavy rainfall and typhoon should be required for the velocity and intensity of them in the regional scales. Many typhoons in western Pacific, western Atlantic oceans, and Indian ocean are generated in the oceanic regions, and then they move toward the large cities with high population density after the time required for prediction of severe weather. However, the earthquakes may occur suddenly and give severe disaster against urban and suburbs without prediction time. Even tsunami gets attack the coastal areas only at several 10 min to hours after large earthquakes. It looks apparently like random phenomena, but the weather phenomena occur following the atmospheric dynamics.

Therefore, the information concerning the large earthquakes should become poorly sufficient for avoiding enough from the severe disaster in the human community. However, the information may be available for avoiding the severe damage by the large earthquakes that the probabilistic forecast of the forthcoming earthquakes includes their time, region, and magnitude. It should be needed for the recovery from the earthquake disaster that there are already founded the safety life line, safety logistic, safety lifeguard, rapid recovery network of community, industry, and speedy re-foundation systems of the soft and hard infrastructure before large earthquakes. These safety network systems should include the natural disaster forecast system near future.

By the way, the next-generation technique which becomes possible to obtain the global seismic activity of the small to micro-earthquakes and also slow slip events by means of new network system determining their source mechanisms and epicenter. The micro- and small earthquakes in the regional areas such as Europe, Japan, and North America are already continuously observing and monitoring and their source data and waveform data are stored in the sophisticated database system for social and scientific use. However, in the global earth system, the earthquakes smaller than magnitude 4 are still not uniform data because of low density configuration of the seismic observation stations in the wide oceanic regions such as Pacific Ocean, Indian Ocean, and Atlantic Ocean. Considering that the spike-like event density of the locality cell covering the global earth of the micro-earthquakes is important and basic information to understand the development of the global mechanics of the surficial solid earth. For the aim of the micro-earthquake observation, the dense global network system of seismic observatory especially in the oceanic regions is quite needed as well as the on-land regions. The possible scenario for this aim is that the ocean-bottom network stations of geophysical observation may be founded along the subduction zones in the global earth as like as the ocean-bottom networks of DONET along the Nankai Trough and NEPTUNE along the Cascadia. These facilities associated with global seismic networks should be very important for data-driven seismological studies including the predictive investigation of large earthquakes in the world together with discovering physical basis of complicated earthquake process in the solid earth in terms of data-driven discovering physics method (Brunton et al. 2016; Iten et al. 2020).

Fin.

Acknowledgements Metadata and data products for this study were accessed through the Northern California Earthquake Data Center (NCEDC), https://doi.org/10.7932/ncedc and International Federation of Digital Seismograph Networks. "NCEDC (2014), Northern California Earthquake Data Center. UC Berkeley Seismological Laboratory. Dataset. https://doi.org/10.7932/ncedc."

The author very thanks to Data Center of Earthquake Research Institute of University of Tokyo for helpful use of cloud application of TSEIS and database of JMA-1. He also thanks much to National Institute of Earthquake Disaster for helpful use of earthquake database.

The author is also very indebted to Professor Yoshiyuki Kaneda of University of Kagawa and Dr. Tatsu Kuwatani of JAMSTEC for many critical discussions and continuous supports for this research. They also give the author many comments of next generation of geophysical researches and their social missions in the near future. He sincerely thanks Prof. H. Sugioka of Kobe University for providing the map of recent large earthquakes in the world.

He is also very grad to Japan Agency of Marine Science and Technology for many supports on writing this book.

Finally, I am deeply indebted to Etsuko Toriumi for many supports and discussions on social efficiency of earthquake disaster and Drs. Taeko Sato, Yosuke Nishida, Gowtham Chakravarthy and Arulmurugan Venkatasalam of Nature-Springer Publisher for many encouragement and comments.

References

Brunton SL, Proctor JL, Kutz JN (2016) Discovering governing equations from data by sparse identification of nonlinear dynamical systems. PNAS 113(15):3932–3937

Iten R, Metger T, Wilning H, del Rio L, Renner R (2020) Discovering physical concepts with neural networks. Phys Rev Lett 124:010508-1–6

Ogata Y (2015) Stochastic prediction of earthquakes—a strategy for the research. Proc Inst Stat Math 63(1):3–27

Satake K, Atwater BF (2007) Long-term perspectives on giant earthquakes and tsunamis at subduction zones. Annu Rev Earth Planet Sci 35:349–374

Appendix A
Application of Recurrent Neural Network (RNN) Modeling for Global Seismicity Dynamics

As discussed in the previous chapters, there are important two conclusions concerning the global correlated seismicity dynamics: One is the logarithmic transformation of local seismicity rates applicable for extracting the feature pattern of their correlated time series by means of orthogonal basis decomposition method (PCA with sparse modeling), and the second is that the correlated seismicity dynamics seems to be expressed by the minimal nonlinear differential equation. Therefore, it is required for inferring dynamical equation from the correlated seismicity rates that the multivariate time series of the global correlated seismicity rates should be investigated by means of nonlinear-type machine learning instead of the linear type state-space modeling (SSM) proposed earlier. On the other hand, the recurrent neural network modeling with long- and short-term memory (LSTM) and convolution neural network model is another candidate for investigation of the multivariate time series of global and regional correlated seismicity dynamics.

As noted in chapters of the data-driven sciences, modeling of the multivariate time series is possibly well-investigated by recurrent neural network modeling with dense layering, long- and short-term memory units (LSTM), and hyperbolic tangent-type activation function (Chollet 2018). Furthermore, it is better to use the basic applications for data-driven sciences that are a combination of Keras and Tensorflow in Python platform. The author now uses them with backend of GPU installed in the cloud computation platform of Google Colaboratory.

There are two objects in applying the deep RNN for multivariate time series analysis: One is prediction of near-future time series from present data series, and the other is to infer the empirical relations among various time series of the multivariate vectors. For examples, considering the correlated seismicity rates in the global earth system, sets of the correlated seismicity rates (z_1, z_2, z_3) must be transformed into subsets of them (z_1, z_2, z_3), (z_1, z_2) and z_1 or the logarithmic seismicity rates in *k*th locality cell, x_k by RNN with LSTM. These available embedding operators may be easily looked for by means of DRNN method. In this Appendix, the author would like to infer these operators in order to clarify the nature of the nonlinear dynamical behavior of the correlated seismicity dynamics in the global and regional seismicity.

M. Toriumi, *Global Seismicity Dynamics and Data-Driven Science*,
Advances in Geological Science,
https://doi.org/10.1007/978-981-15-5109-3

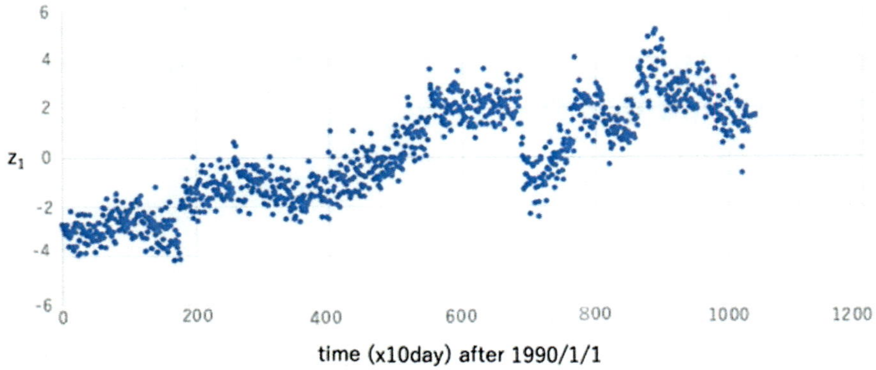

Fig. A.1 RNN regression time series of the global correlated seismicity z_1 by means of simple RNN, showing similar variation pattern with raw z_1

The transformation from the correlated seismicity rates, z_1 to the local cell seismicity rates is also tried for direct prediction of the seismicity of the individual locality using DRNN modeling.

First, the author intends to study the relationship between original z_1 time series and the time series processed by training the original z_1 (called as RNN regression in this book) in DRNN. Figure A.1 shows the RNN transformation of z_1 time series, showing clearly the regression results of the z_1 time series. Here, the time step for RNN transformation is 5 steps as the input data and output data from DRNN are the single one; that is, the set of the training data is composed of five input serial data and target of next one data in the time series as shown in Fig. A.1. The one-step-ahead prediction of the correlated seismicity rate can be obtained using the weight coefficients determined by many training and testing repeat process. Actually, the prediction for every 50 steps ahead has been performed in the case of the single time series z_1 in the global earth as shown in Fig. A.2. It shows the gradual extension from the observed time series of z_1 to the 500 days ahead as same as predicted in the SSM modeling in the previous chapter.

Next, the author tried to carry out the training and testing of input data separated early and late segments of the time series of z_1 as shown in Fig. A.2. This means the check of robustness of the parameters set in RNN with LSTM regression of the time series. Considering that the data cloud defined in the text keeps invariance against time advancement, the so-called cross-validation by means of splitting the time series into several segments is possibly applied to validation of DRNN modeling of the time series.

It is very important that the DRNN transformation of z_1 both by simple RNN and LSTM RNN makes available parameter sets for the later z_1 time series, because the z_1 in the later half time series determined by parameter sets of early half series of z_1 shows similar variation compared RNN regression z_1. Thus, it seems to be possible that the time series of z_1 in the whole range can be applied for prediction of z_1 in

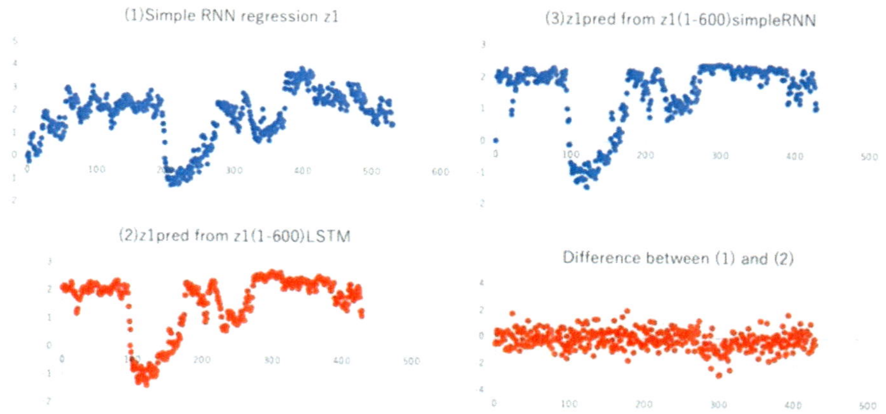

Fig. A.2 Comparison of testing patterns of late half time series of z_1 from RNN early half training patterns by simple RNN and LSTM, and difference between regression and testing z_1 showing the abnormal residual in the period between 280 and 350 steps

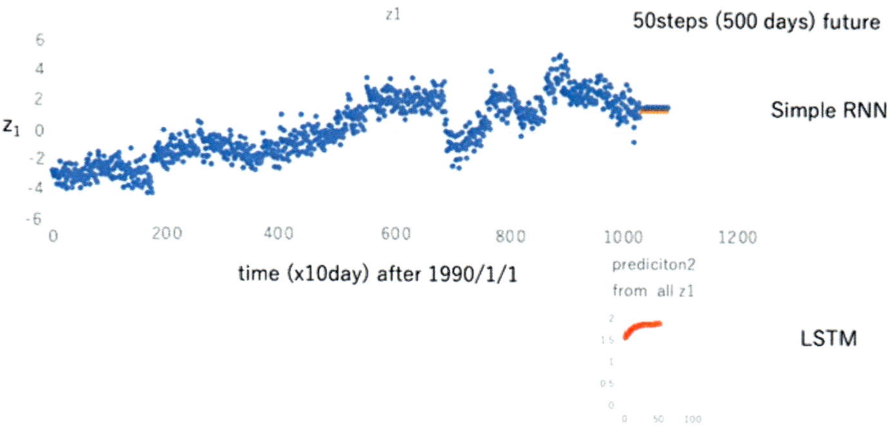

Fig. A.3 Prediction of 10 and 50 steps ahead by simple RNN (upper) and LSTM (lower) modeling of the global correlated seismicity rate z1, showing the similar prediction trends

the near-future seismic activity. In Fig. A.3, the predicted z_1 of the global seismicity is shown for values until 10 steps ahead and 50 steps ahead. The prediction was performed using the last five steps segments of the time series and successive five steps adding the one-step-ahead prediction value for next prediction. So far as the data cloud maintains the features even in the 50 steps ahead, the prediction from the known time series of z_1 to the near future may be applicable for the global and regional seismicity.

As discussed in the prediction by SSM modeling, the prediction of z_1, z_2, and z_3 in near future is possible and their combination makes possible to predict also the real

Fig. A.4 Regression time series of logarithmic seismicity rates x by LSTM and relations of x and the correlated seismicity rate z_1 for Chile (top) and Northeast Japan (middle), Sumatra and Kermadec (bottom) districts in the global seismicity regime

seismicity rates of the individual locality such as northeastern Japan region (NEJ) under the recognition of the pseudoinverse function from z to the local logarithmic seismicity rates x. The pseudoinverse function from z to x is possibly inferred by the transformation of z to x using the DRNN LSTM modeling as shown in Fig. A.4. The relationships between z_1 and x by RNN regression from z_1 of the cell locality such as Chile, Sumatra, Tonga–Kermadec, and Northeast Japan are shown, and those display somewhat nonlinear relations. It looks like that the relations between them commonly are constituted with two segments having different slopes. Furthermore,

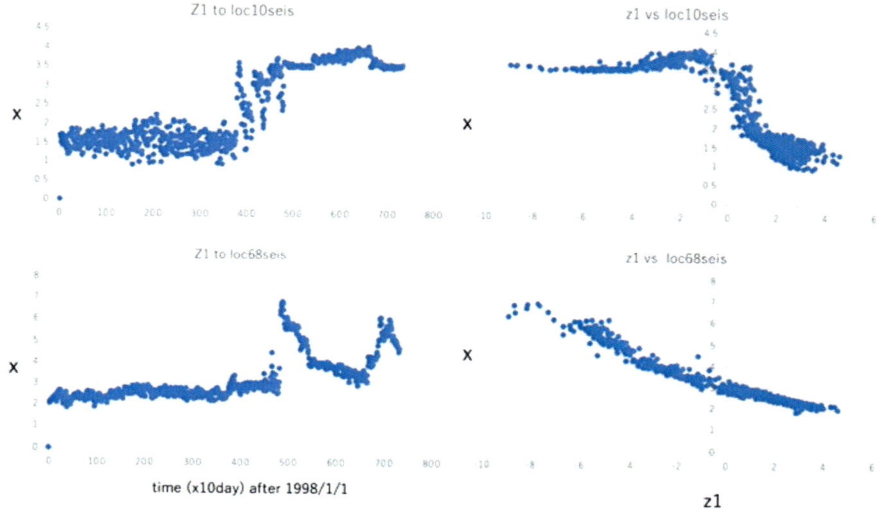

Fig. A.5 Regression time series of logarithmic seismicity rates x by simple RNN and the relations of x and z1 for Okinawa (top) and Northeast Japan (bottom) districts in the Japanese islands region system

it seems possible that the machine learning with DRNN of correlated seismicity rates z and local seismicity rates is available for the prediction of x of the local seismicity rates in the near future from the correlated seismicity rate z_1. The relations between z_1 and the RNN-LSTM transformed logarithmic seismicity rate x as shown in Fig. A.5 for Okinawa and Northeast Japan cells, suggesting the clear two segments in Okinawa and simple near-linear relation of Northeast Japan cell. It seems that the correlated seismicity with RNN with LSTM modeling is very available for predictive study of the seismicity rates of the locality cells in the global and regional seismicity dynamics.

Reference

Chollet F (2017) Deep learning with Python. Manning

Appendix B
Comments on Databases and Software Used in This Book

The author developed database stored in Electronic Supplement Materials (ESM_1tori ~ESM_20tori) of this book of the micro- and small event density of earthquake in the local cells along the global subduction zones and Japanese islands arc and Northern California regions, based on the database of USGS data center, University of California, Berkeley, ERI of University of Tokyo, Japan Meteoric Agency (JMA), and National Institute of Earthquake Disaster (NIED). The author thanks deeply these institutions. The new database developed by the author includes the lists of longitude and latitude of the locality cells in the global earth and regions of Japanese islands and Northern California. The time series of the seismicity rates and the locality of the cells studied in the book are also opened as the database from 1990 to 2018 in the global earth and 1990–2019 in the Northern California region, and 1998–2017 or 2018 in the Japanese Islands region.

The application software used in this book is as follows:

Mac OS 10.15, Windows 10,
RStudio1.3, R packages; KFAS, sparsepca, pca, mclust, fastICA,
Anaconda3, python3, keras, tensorflow, numpy, pandas, sciKit, matplot,
Google Colaboratory, python 3.0, keras, tensorflow 1.14,
The author thanks sincerely these applications builders.

M. Toriumi, *Global Seismicity Dynamics and Data-Driven Science*, Advances in Geological Science,
https://doi.org/10.1007/978-981-15-5109-3